pulsating stars

pulsating stars

a NATURE reprint

with introductions by
Professor F. G. Smith &
Dr A. Hewish

Springer Science+Business Media, LLC

FISHER KNIGHT AND CO LTD
Lattimore Road St Albans

Contents

Optical Measurements

Theories

Applications

Introduction

The Pulsating Stars

by

F. G. SMITH

University of Manchester,

Nuffield Radio Astronomy Laboratories,

Jodrell Bank

THE discovery of the pulsating radio stars by Miss Jocelyn Bell, working with Dr Hewish at Cambridge, rivals the early classic discoveries of extraterrestrial radio sources in its importance and in its excitement. It may at first have seemed to be a lucky accident that the pulsars, as they are now called, should be detected first in an experiment designed to investigate the solar wind; the fact that for several months after the publication of the discovery no other pulsars had been detected in any other radio observatory shows that the discovery followed naturally on the many years of work that Hewish has put into the study of another type of fluctuating signal, the interplanetary scintillation of discrete radio sources.

The announcement of the discovery of the first pulsating radio source was made in February 1968 (paper 1) followed by the details of the three others in April (2). The excitement spread rapidly among observers and theorists alike, and their work found a ready forum in *Nature* where in the next 5 months more than forty papers were published. Two weeks after the first announcement an article from Jodrell Bank (6) described observations of *CP* 1919 over a wide range of radio frequencies, giving pulse shapes, an average pulse profile, and a description of the remarkable fading pattern of the signal strength. The flood of theory started with two papers (42, 39) only three weeks later.

At this time (August 1968) almost the whole of the literature is to be found in *Nature*, and it is the intention of the next few pages to provide a guide to these papers in the form of a discussion of the main topics under the headings of Observational Techniques, Pulse Characteristics, Radio Frequency Spectrum, Fluctuations, Periodicity, Theories of the Periodicity, Optical Searches, Emission Theory and the Interstellar Medium.

Observational Techniques

The average radio power flux from a pulsar is very small compared with most cosmic radio sources. The periodicity of the pulses is, however, remarkably constant, and once the period is known, the detection of a pulsar is comparatively easy. Observations of the detailed characteristics of the radio pulses have been made mainly with the large steerable radio telescopes at Jodrell Bank (6, 7, 10, 11, 12), the 210 ft radio telescope at Goldstone (9), the 210 ft at Parkes (17, 15), and more recently with the 300 ft transit telescope at Green Bank. At Cambridge, besides the "interplane-tary scintillation" array, observations have been made with a parabolic trough telescope the beam of which can track in right ascension (8), and accurate positions have also been measured with the One Mile telescope (19, 2, 21). The large "Northern Cross" at Bologna was used to find the declinations of three of the pulsars (13). Observations from Arecibo have been directed at the fine structure of the pulses (ref. 14, and in *Science*, **160**, 758; 1968).

With a large radio telescope it is possible to detect individual pulses and to measure their detailed shapes (as, for example, in ref. 6), but great sensitivity can only be achieved by the superposition of many pulses in some form of on-line computation. This has been achieved for both optical and radio observations either by adopting multichannel signal or pulse-height analysers (for example, in refs. 9, 25, 15) or by using a general-purpose digital computer (6). The search for further pulsars contains a challenge to find such analytic methods which will apply to a signal known to have a precise period, but whose period is unknown to within a factor of ten or more.

Pulse Characteristics

The detailed shapes and amplitudes of the radio pulses are very variable. A series of pulses from *CP* 1919 contains fluctuating components down to 1 ms duration (6) contained in a total pulse duration up to 50 ms. Components down to 0·2 ms have been found in pulses from *CP* 0950 (14). Individual components seem to persist from pulse to pulse in *CP* 1919 and in *CP* 1133 (10).

An average of 100 pulses shows a much more regular behaviour; the "pulse profiles" obtained in these averages are typical of the individual pulsars and the same over a wide range of frequencies (9, 10).

The radiation was at first thought to be confined to a single pulse within a profile lasting about 50 ms for all four pulsars, but for *CP* 0950 an "interpulse" was found to occur 100 ms before the main pulse (11). The pulse energy is only about 1½ per cent of that of the main peak, and it can usually be detected only by integration over several minutes.

It was suggested that alternate pulses of *CP* 1133 have different amplitudes (13), but this has not appeared in other observations in spite of searches prompted by theories based on models of binary stars.

Linear polarization of the radio pulse was discovered in *CP* 0950, where it reached 95 per cent (7). Other sources showed less polarization, but it was found (7, 14, 15) that separate components of the pulses were highly polarized, not all in the same directions. Circular polarization was discovered at Arecibo (14) and at the National Radio Astronomy Observatory, and it now seems that the polarization is usually elliptical and also that it varies over a period of days (24).

Spectrum

The initial detection at 81 MHz was followed by measurements at frequencies up to 1,410 MHz (6, 10, 15), but it was clear that the flux density fell markedly at higher frequencies. From successful observations at 2,295 and 2,695 MHz (9, 12, 15) the average spectrum was found to show a spectral index $\alpha = 1.5$ for CP 0950 and CP 1133, a steeper slope for CP 0834, and $\alpha > 3$ for CP 1919 at frequencies above 1,000 MHz. (Spectral index is defined by the relation flux density $\propto$ frequency$^{-\alpha}$). The spectrum is difficult to define precisely, as the pulse strength fades deeply on all frequencies.

Fluctuations

CP 1919 did not appear on all the records in the series searched by Miss Bell, and on those where it did appear the strength was variable from pulse to pulse. Variations are now known to occur on a wide range of time scales, from pulse to pulse, and over minutes, hours and days (6, 9, 10, 13, 15).

From pulse to pulse there is some correlation in strength over a series of up to 20 pulses for CP 1919 (8, 10), but no such correlation is found for the other three. The pulse energy averaged over several minutes shows variations of 10 : 1 during a period of a few hours at 408 MHz and higher frequencies (9, 10), with a probability distribution containing many more high values than would be expected from a Rayleigh distribution. At lower frequencies the fluctuations seem to be more rapid, and more nearly Rayleigh-like; even so, there are slow variations from day to day which have not yet been followed in detail.

The fading pattern of pulse energy becomes unrelated at radio frequencies differing by only a small amount. CP 0950 shows such frequency structure within a bandwidth of 1 MHz at 81 MHz (8); the observations of polarization at 151 MHz displayed the same structure within 4 MHz, although it was at first attributed to gain changes in the receivers (47). No decorrelation is observed over 1 MHz at 430 MHz for CP 0950 (14). CP 1919 shows decorrelation within 1 MHz at 151 MHz (15), and a similar result has been reported for CP 1133 (24).

A wide bandwidth has, however, been found for some detailed variations from pulse to pulse (10), and it is thought that this variation at least is intrinsic to the source, leaving the deep fading to be explained by some form of scintillation. The variations from pulse to pulse have been found to be identical and simultaneous over a baseline of 325 km (16), and that they are not due to interstellar scintillations.

Measurement of Periodicity

In their early measurements the Cambridge observers found an accurate period for CP 1919, but apparently missed count of one pulse per day (9, 17, 18, 24). Heliocentric periods accurate to 1 part in 10^7 are available (*IAU Circular No.* 2072), and there seems a

good chance of improving this to 1 part in 10^9, as a group of pulses can be timed to an accuracy of a few milliseconds.

So far no departures from a regular periodicity have been reported. Apart from a possible secular change of period to be expected from changes in the source, there might be expected to be a Doppler effect due to a slow orbiting of the source round another star. If the source produces a period accurate to better than 1 part in 10^9, there is a chance of observing the general relativistic effect due to the eccentric motion of the Earth round the Sun (50), or even due to an eccentric orbit in the source (51).

Theories of the Periodicity

A periodicity accurate to at least 1 part in 10^7 suggests a massive dynamical system, such as a pulsating star, a rotating star, or a binary star. It has, however, been pointed out that the period of about 1 s is also close to the time scale of a supernova explosion (27), and to the free fall time of a shell of material round an unstable white dwarf (33, 34).

Radial oscillations of a white dwarf had already been discussed by Meltzer and Thorne (*Ap. J.*, **145**, 514; 1966), but their theoretical work indicated that fundamental modes of oscillation could only have periods greater than 8 s. A succession of papers (39, 30, 28, 31, 38) investigated more exact models, in which the material was predominantly helium, carbon, iron or even heavier elements, and fundamental periods down to 2 s seem now to be possible. Rotation shortens the period still more, and if differential rotation occurs the period for a white dwarf composed of heavy elements could be as short as 0·1 s (29, 32). Neutron stars would oscillate radially with periods of the order of 1 ms, but they may become partly crystalline, allowing a torsional oscillation with a period nearer to 1 s (35).

The oscillation of a white dwarf might be expected to be in an overtone mode, as the maximum amplitude then occurs at the surface, where the oscillation could be maintained by a shell of burning hydrogen. An energy source could, however, exist within the interior, provided it was a low temperature reaction in heavy elements at high pressure (28). This difference is most important in the question of optical identification, because a white dwarf hot enough for hydrogen burning should be visible at a considerable distance.

The rotating star hypothesis, in which the emission is beamed from a single spot on the surface (39, 40), cannot easily explain the accurate timing of the onset of the pulse and the details of pulse shape and average profiles. A binary system was suggested, first with each of two condensed stars acting as a gravitational lens (42), then a neutron star with a satellite, analogous to the Jupiter Io system (43). All such condensed binary systems suffer from the rapid loss of energy by gravitational radiation, and from the disruption of a satellite by tidal forces (44, 45).

The pulsating white dwarf seems at present to be the only reasonable contender still holding the field, and it was to be expected that a sufficiently accurate measurement of position would provide an identification with a recognizable white dwarf star.

Optical Searches

Identification with the faintest objects on the 48 inch Sky Survey plates requires a position accurate to about 100 square seconds of arc. An accuracy approaching this is available (19, 20, 13, 21, 22) for a few of the pulsars, but no optical object brighter than magnitude 20·5 can be seen in any of these positions. At one time a less accurate position for *CP* 1919 allowed the possibility that a star of magnitude 17·5 might be the correct identification, and in several observatories a search was made for periodic fluctuations of light from this star or from the sky around it (23, 25). No reliable positive result has been obtained, although preliminary observations at Kitt Peak and at Lick Observatory held out some promise (24).

According to the measurements of group delay in the pulses *CP* 0950 should be the closest of the pulsars, and its distance has been estimated at 30 pc (20, 10). If this distance is assumed to be correct, the optical luminosity must be less than $5·4 \times 10^{-6}$ that of the Sun, that is, the star must be fainter than absolute luminosity $M = 18$, and outside the usual range of white dwarfs (21). If the white dwarf theory is to be maintained, then either the distances must be greater or the stars must be cooler than the usual white dwarfs with spectral classification *Bo–Ko*.

Emission Theory

Because the radio pulse leaves the source over a wide range of frequencies within a few milliseconds (6, 10), the signal strength at the source must be very high indeed. If the source in *CP* 1919 is the whole area of a white dwarf at a distance of 60 pc, the pulse power is 10^{21} W, the pulse energy is 2×10^{19} J, and the flux density at the surface is 3×10^{5} W m^{-2}, corresponding to an electric field strength of 10^{4} V m^{-1} (6). The source of the radiation is probably very much smaller than such an area, possibly only 60 km across (14), giving even larger values of flux density. The brightness temperature exceeds 10^{21} °K, and all possibility of incoherent radiation mechanism is therefore excluded (19, 6, 7). For example, the attempt to use synchrotron radiation theory (33) was totally unable to account for the high intensities. A coherent mechanism, possibly like that operating in the radio pulse from a cosmic ray shower (7), is essential, but the different radio frequencies could be emitted not precisely simultaneously but by the excitation in rapid succession of a series of resonant regions, as in a solar radio outburst. The suggestion that the observed pulse delay is due to such a succession (36) does not, however, account for the accurate delay law.

A coherent oscillation is expected to be polarized, but no detailed explanation is available for the range of polarizations observed. It has been suggested that the generation is basically in a circular mode, and that mode coupling subsequently converts part of the radiation to linear polarization (18).

Rotators (and orbiters) provide ready explanations for the conversion of energy from mechanical modes to radiation, via the interaction of a rotating magnetic field with a plasma cloud (40, 41), but again no precise mechanism has been suggested.

Leaving aside the unsolved question of the actual mechanism of emission, analysis of the average profile of the pulses may give a clue to the size and geometry of the emitting star. The original discovery paper (1) suggested that the pulse length might be related to the light travel time across the radius of a star, and the subsequent discovery that the "interpulse" from *CP*

0950 was polarized differently from the main pulse suggested that it originated in a different region excited at an earlier time in the pulse period (7). Following these remarks, analysis of the pulse profiles in ref. 10 has led to a model in which the emission is confined largely to two zones of latitude in a white dwarf star, with excitation starting at the poles and moving toward the equator (37). The model gives an estimate of radii near 7,000 km for *CP* 0950 and near 14,000 km for the other three pulsars. These agree well with the radii of known white dwarf stars.

Again the theoretical work inclines toward the white dwarf interpretation, but by no means proves it.

The Interstellar Medium

The dispersion in the arrival time of the radio pulses over the radio frequency band from 2,700 MHz to 81 MHz fits very precisely the law expected from delay in ionized interstellar gas (6, 10). Values of total electron content in the line of sight lead to estimates of distance between 30 pc (*CP* 0950) and 128 pc (*CP* 0834), on the assumption that the average electron density in interstellar space is 0·1 cm^{-3}. If the distances can ever be measured by some other means, this calculation could be inverted to provide a much more accurate measurement of the electron density.

The fluctuations in pulse strength may be due to scintillation in irregularities in the interstellar gas, or in a gas cloud close to the source. The probability distribution of the fluctuations is, however, unlike the Rayleigh distribution typical of interplanetary scintillation; furthermore, the frequency dependence and the coherence bandwidth are such that two scintillation mechanisms appear to be operating simultaneously (46). A further study of the form of the variations and their dependence on frequency may show which part of these fluctuations is generated within the source, how much is scintillation close to the source, and how much is in the interstellar medium.

Faraday rotation of the plane polarized components of the radio pulse offers a remarkable method for measuring the interstellar magnetic field. The group delay measures the electron content, and the rotation measures the product of field and electron content. An upper limit of 2×10^{-7} gauss for the average field was obtained in the first measurement on *CP* 0950 (47, 48, 49), a very much smaller value than was expected. Further measurements may be possible when the polarization structure of the pulses is better understood.

The Next Stage

Any suspicions that pulsars would be a nine-days wonder must by now have been dispelled by a glance at the diverse topics of the collected papers. It would, of course, have been encouraging to find a visible object in one of the accurately determined positions, but the chance of finding the type of the pulsating stars through their visible spectra now seems very slight. As compensation, it may become clearer that the star is one which cannot be reached optically, so that the radio pulses offer the only means of detection of an object the physical processes of which are of great interest. The theory of the internal constitution of a star which can not only resonate at a frequency of about 1 Hz but which contains a driving mechanism to sustain such an oscillation scarcely exists. The place of the star in the evolutionary sequence, and the contribution of its mass to the general composition of the galaxy, both raise extremely interesting problems.

If radio methods must stand alone, much can still be done to fill out the very tentative picture of the pulsed excitation of the surface of a white dwarf star, which emerges as the most likely theory so far. Possibly this picture can be improved so as to give details such as ellipticity and rotation rate, and the configuration of the magnetic field; these would emerge from further study of the pulse shapes and polarizations. The great accuracy with which the periodicity can be measured suggests that a secular change might be detected, which could be attributed to a progressive change of constitution or a loss of energy even over only a few years.

For the interstellar medium, a long series of observations will be needed before the fading pattern can be understood in terms of scintillation in ionized clouds. The magnetic field may remain an open question if it turns out that the plane polarization which appeared so fortunately in *CP* 0950 is not sufficient for the measurement to be made in other pulsars. The possibility of making the first such direct measurement is so attractive, however, that this experiment will surely be repeated many times over.

Finally, the search for more pulsars is going ahead in several observatories. The first few were discovered without the help of sophisticated computation techniques, in a survey covering less than half the sky. The fifth to be discovered, by Harvard observers at the National Radio Astronomy Observatory at Green Bank (4), was found by computer analysis of a tape recording, and this technique is being applied also at Arecibo. Two more pulsars have been found by Cambridge observers when the search was extended to higher latitudes (3). Two have been found in the Southern Hemisphere (5); these are distinguished one by its proximity to the direction of the centre of the galaxy, the other by its long period, 1·96 s.

The total of nine by early September may look like slow progress, and indeed it is clear that the radio observers are finding their task is not an easy one. The realization of the difficulties in finding more pulsars must surely enhance the credit already given to Miss Bell and Dr Hewish for their wonderful discovery.

The Discovery of Pulsars

by
A. HEWISH
Mullard Radio Astronomy Observatory,
Cavendish Laboratory,
University of Cambridge

Now that pulsars have become firmly established as one of the most intriguing phenomena in astrophysics today it may well be asked why were they not discovered sooner? After all, the era of powerful radio telescopes began at least ten years ago. The principal reason is simply that pulsars are extremely weak radio sources and their mean signal strength is such as to place them below the detection limit of the large-scale sky surveys which have so far been made. Add to this the fact that they are sporadic emitters, often requiring repeated observations to be sure of their presence, and that modern radio telescopes have a progressive tendency to operate at ever shorter wavelengths where the pulsar spectrum falls away, and the relatively late appearance of pulsars becomes easier to understand.

The stage was set, however, by the discovery in 1964 of a phenomenon known as interplanetary scintillation. This effect, which only occurs for radio sources of exceedingly small angular diameter, such as quasars, takes the form of a rapid fluctuation of intensity on a time-scale of the order of one second. It is induced by the diffraction of incoming radio waves as they traverse plasma clouds in interplanetary space and its presence gives a most valuable indication of the angular size of a radio source.

A radio telescope designed specifically to exploit interplanetary scintillation was completed at Cambridge during 1967. Because scintillation is a plasma effect it is less pronounced at short wavelengths and so the instrument used the relatively long wavelength of 3·7 metres (81·5 MHz). To achieve an adequate sensitivity a receiving aerial with a collecting area of $4\frac{1}{2}$ acres was necessary and giant structures of this size cannot, of course, be made fully steerable. The antenna took the form of a rectangular array containing 2,048 dipoles, the reception beam of which could be steered in declination by phase-scanning. Earth rotation was used to scan in right ascension which meant that a given source was in transit for about three to four minutes.

Routine survey recordings began in July 1967 and a considerable area of sky came under repeated surveillance at a high sensitivity for the first time in the history of radio astronomy. One basic advantage of the phase-scanning technique, as compared with a steerable radio telescope, is that of multiple-beam operation which enables several different regions of the sky to be observed simultaneously, thereby reducing the survey time. Use of this technique made it possible to scan the whole sky in the range of declination $-08° < \delta < 44°$ within one week and repetitive observations on this basis were undertaken. Naturally, in view of the scintillation data being sought, the recording techniques were designed to be particularly sensitive to radio sources exhibiting a fluctuating intensity and thus, quite fortuitously, the conditions for detecting pulsars were ideally fulfilled.

The data from the survey consisted of pen recordings which accumulated at some 400 feet per week, and part of the analysis carried out by Miss Bell was a plot, in celestial co-ordinates, of the position of any scintillating component of received intensity. Towards the end of August she had plotted, on more than one occasion, what appeared to be a scintillating source which was conspicuous because it was in transit near midnight at a time when interplanetary scintillation falls to a low value. On account of the sporadic nature of the source its right ascension appeared to shift irregularly and so we took it to be some kind of interference. By the end of September, however, it was clear that the right ascension was, indeed, constant and that the apparent shifts in position were explained by the random occurrence of the signals for a period shorter than the drift-time of the source through the reception beam. The absence of measurable parallax then showed that the source must be located far outside the solar system and the possibility of radio interference from a deep space probe was eliminated.

At this time it was imagined that the source might be some kind of radio flare star and further investigations were planned to see if the signals had characteristics similar to radio outbursts from the Sun. During October and November the intensity fell to a low value and it was not until November 28 that the first satisfactory observation was made. Pulses could be distinguished on this recording, which may be said to mark the discovery of pulsars, but it was only several days later, when the pulses had been confirmed, that the results were taken to be genuine. Signals of this kind are just not expected from a celestial source and a certain scepticism was inevitable.

It was immediately evident from the short duration of the pulses that the source must be exceedingly small. But it was not until the frequency drift arising from dispersion had been found that Dr Pilkington was able to estimate the true duration of the emitted pulse to be of the order of 20 ms. This showed that the source was of planetary dimensions and the possibility that the signals might be from an extra-terrestrial civilization could no longer be ignored. The rapid variation of pulse amplitude was suggestive of some type of code, but extended observations made by tracking the source with a 470 m × 20 m reflector normally used by Dr Scott and Mr Collins in a lunar occultation programme gave no indication of a recognizable pattern.

Accurate measurements of the pulse repetition frequency were carried out to investigate the feasibility of detecting planetary motion of the source by virtue of the Doppler effect, and the astonishing regularity of the pulses was then found. Early in January 1968 it was clear that the heliocentric pulse rate was constant to better than one in 10^7; a Doppler variation was present but this was entirely accounted for by the Earth's orbital motion and it therefore seemed highly unlikely that the source could be a planet. A systematic search was then undertaken for further pulsed sources. This consisted of a re-examination of the three miles of survey charts by Miss Bell, followed by additional observations of sources which seemed to be possible pulsars. Within a few weeks three more pulsars had been located in widely differing parts of the sky and this confirmed our opinion that pulsars must be a natural phenomenon.

In these preliminary observations over a small range of wavelength it was not certain whether the frequency drift was caused by dispersion in the interstellar medium or in the source itself, but it was possible to place an upper limit to the pulsar distance based on an assumed interstellar electron content. This showed that the pulsars must be local objects within the galaxy. The only plausible candidates to account for pulsars, having regard to their small physical size, seemed to be white dwarf stars or the hypothetical neutron stars. We suggested that radial vibration of the entire star afforded the simplest explanation of the constant period and that shock waves in the atmosphere might initiate radio flashes by some high energy particle-plasma interaction. The difficulty of accounting for the observed period was, however, pointed out—the white dwarf periods being rather too long and those of the neutron star rather too short.

These results were submitted for publication eight weeks after the pulsed nature of the radiation had first been recognized. Opinion has been expressed that this represented an undue delay. Bearing in mind the importance of obtaining evidence for or against the planetary hypothesis, and having regard to the publicity which any statement at this stage would undoubtedly have provoked, it was felt premature to release information until the Doppler measurements had yielded positive results.

Now, nearly one year later, many new and exciting results have been obtained from radio observatories throughout the world. At the time of writing ten pulsars have been reported—six from Cambridge, two from the USA and two in the southern sky from Australia. The number will undoubtedly grow as more advanced search techniques become available. Speculation about the nature of the sources has been rife and vivid, and it is encouraging that white dwarf vibration still offers a plausible explanation of these mysterious and fascinating bodies.

Discovery

1. Observation of a Rapidly Pulsating Radio Source

by

A. HEWISH
S. J. BELL
J. D. H. PILKINGTON
P. F. SCOTT
R. A. COLLINS

Mullard Radio Astronomy Observatory,
Cavendish Laboratory,
University of Cambridge

In July 1967, a large radio telescope operating at a frequency of 81·5 MHz was brought into use at the Mullard Radio Astronomy Observatory. This instrument was designed to investigate the angular structure of compact radio sources by observing the scintillation caused by the irregular structure of the interplanetary medium[1]. The initial survey includes the whole sky in the declination range $-08° < \delta < 44°$ and this area is scanned once a week. A large fraction of the sky is thus under regular surveillance. Soon after the instrument was brought into operation it was noticed that signals which appeared at first to be weak sporadic interference were repeatedly observed at a fixed declination and right ascension; this result showed that the source could not be terrestrial in origin.

Systematic investigations were started in November and high speed records showed that the signals, when present, consisted of a series of pulses each lasting $\sim 0·3$ s and with a repetition period of about 1·337 s which was soon found to be maintained with extreme accuracy. Further observations have shown that the true period is constant to better than 1 part in 10^7 although there is a systematic variation which can be ascribed to the orbital motion of the Earth. The impulsive nature of the recorded signals is caused by the periodic passage of a signal of descending frequency through the 1 MHz pass band of the receiver.

The remarkable nature of these signals at first suggested an origin in terms of man-made transmissions which might arise from deep space probes, planetary radar or the reflexion of terrestrial signals from the Moon. None of these interpretations can, however, be accepted because the absence of any parallax shows that the source lies far outside the solar system. A preliminary search for further pulsating sources has already revealed the presence of three others having remarkably similar properties which suggests that this type of source may be relatively common at a low flux density. A tentative explanation of these unusual sources in terms of the stable oscillations of white dwarf or neutron stars is proposed.

Position and Flux Density

The aerial consists of a rectangular array containing 2,048 full-wave dipoles arranged in sixteen rows of 128 elements. Each row is 470 m long in an E.–W. direction and the N.–S. extent of the array is 45 m. Phase-scanning is employed to direct the reception pattern in declination and four receivers are used so that four different declinations may be observed simultaneously. Phase-switching receivers are employed and the two halves of the aerial are combined as an E.–W. interferometer. Each row of dipole elements is backed by a tilted reflecting screen so that maximum sensitivity is obtained at a declination of approximately $+30°$, the overall sensitivity being reduced by more than one-half when the beam is scanned to declinations above $+90°$ and below $-5°$. The beamwidth of the array to half intensity is about $\pm \frac{1}{2}°$ in right ascension and $\pm 3°$ in declination; the phasing arrangement is designed to produce beams at roughly $3°$ intervals in declination. The receivers have a bandwidth of 1 MHz centred at a frequency of 81·5 MHz and routine recordings are made with a time constant of 0·1 s; the r.m.s. noise fluctuations correspond to a flux density of $0·5 \times 10^{-26}$ W m^{-2} Hz^{-1}. For detailed studies of the pulsating source a time constant of 0·05 s was usually employed and the signals were displayed on a multi-channel 'Rapidgraph' pen recorder with a time constant of 0·03 s. Accurate timing of the pulses was achieved by recording second pips derived from the *MSF* Rugby time transmissions.

A record obtained when the pulsating source was unusually strong is shown in Fig. 1a. This clearly displays the regular periodicity and also the characteristic irregular variation of pulse amplitude. On this occasion the largest pulses approached a peak flux density (averaged over the 1 MHz pass band) of 20×10^{-26} W m^{-2} Hz^{-1}, although the mean flux density integrated over one minute only amounted to approximately $1·0 \times 10^{-26}$ W m^{-2} Hz^{-1}. On a more typical occasion the integrated flux density would be several times smaller than this value. It is therefore not surprising that the source has not been detected in the past, for the integrated flux density falls well below the limit of previous surveys at metre wavelengths.

The position of the source in right ascension is readily obtained from an accurate measurement of the "crossover" points of the interference pattern on those occasions

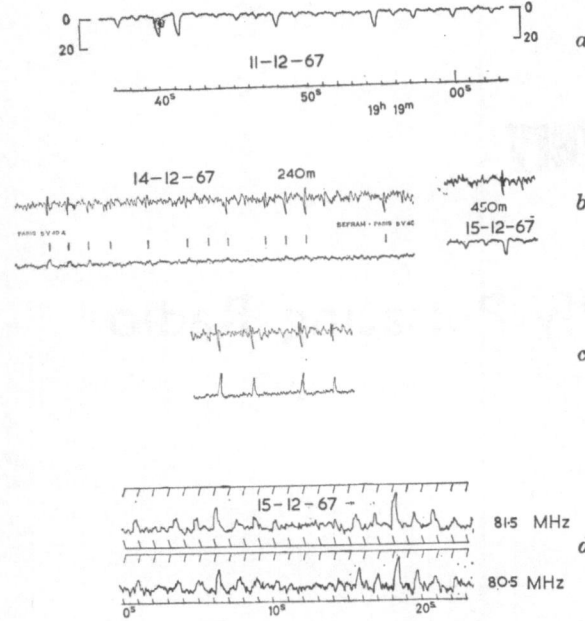

Fig. 1. *a*, A record of the pulsating radio source in strong signal conditions (receiver time constant 0·1 s). Full scale deflexion corresponds to 20×10^{-26} W m⁻² Hz⁻¹. *b*, Upper trace: records obtained with additional paths (240 m and 450 m) in one side of the interferometer. Lower trace: normal interferometer records. (The pulses are small for $l = 240$ m because they occurred near a null in the interference pattern; this modifies the phase but not the amplitude of the oscillatory response on the upper trace.) *c*, Simulated pulses obtained using a signal generator. *d*, Simultaneous reception of pulses using identical receivers tuned to different frequencies. Pulses at the lower frequency are delayed by about 0·2 s.

when the pulses were strong throughout an interval embracing such a point. The collimation error of the instrument was determined from a similar measurement on the neighbouring source 3C 409 which transits about 52 min later. On the routine recordings which first revealed the source the reading accuracy was only ± 10 s and the earliest record suitable for position measurement was obtained on August 13, 1967. This and all subsequent measurements agree within the error limits. The position in declination is not so well determined and relies on the relative amplitudes of the signals obtained when the reception pattern is centred on declinations of 20°, 23° and 26°. Combining the measurements yields a position

$$\alpha_{1950} = 19\text{h } 19\text{m } 38\text{s} \pm 3\text{s}$$

$$\delta_{1950} = 22° \, 00' \pm 30'$$

As discussed here, the measurement of the Doppler shift in the observed frequency of the pulses due to the Earth's orbital motion provides an alternative estimate of the declination. Observations throughout one year should yield an accuracy of ± 1'. The value currently attained from observations during December–January is $\delta = 21° 58' \pm 30'$, a figure consistent with the previous measurement.

Time Variations

It was mentioned earlier that the signals vary considerably in strength from day to day and, typically, they are only present for about 1 min, which may occur quite randomly within the 4 min interval permitted by the reception pattern. In addition, as shown in Fig. 1*a*, the pulse amplitude may vary considerably on a time-scale of seconds. The pulse to pulse variations may possibly be explained in terms of interplanetary scintillation[1], but

this cannot account for the minute to minute variation of mean pulse amplitude. Continuous observations over periods of 30 min have been made by tracking the source with an E.–W. phased array in a 470 m × 20 m reflector normally used for a lunar occultation programme. The peak pulse amplitude averaged over ten successive pulses for a period of 30 min is shown in Fig. 2*a*. This plot suggests the possibility of periodicities of a few minutes duration, but a correlation analysis yields no significant result. If the signals were linearly polarized, Faraday rotation in the ionosphere might cause the random variations, but the form of the curve does not seem compatible with this mechanism. The day to day variations since the source was first detected are shown in Fig. 2*b*. In this analysis the daily value plotted is the peak flux density of the greatest pulse. Again the variation from day to day is irregular and no systematic changes are clearly evident, although there is a suggestion that the source was significantly weaker during October to November. It therefore appears that, despite the regular occurrence of the pulses, the magnitude of the power emitted exhibits variations over long and short periods.

Instantaneous Bandwidth and Frequency Drift

Two different experiments have shown that the pulses are caused by a narrow-band signal of descending frequency sweeping through the 1 MHz band of the receiver. In the first, two identical receivers were used, tuned to frequencies of 80·5 MHz and 81·5 MHz. Fig. 1*d*, which illustrates a record made with this system, shows that the lower frequency pulses are delayed by about 0·2 s. This corresponds to a frequency drift of ~ -5 MHz s⁻¹. In the second method a time delay was introduced into the signals reaching the receiver from one-half of the aerial by incorporating an extra cable of known length l. This cable introduces a phase shift proportional to frequency so that, for a signal the coherence length of which exceeds l, the output of the receiver will oscillate with period

$$t_0 = \frac{c}{l}\left(\frac{\text{d}\nu}{\text{d}t}\right)^{-1}$$

where $\text{d}\nu/\text{d}t$ is the rate of change of signal frequency. Records obtained with $l = 240$ m and 450 m are shown in Fig. 1*b* together with a simultaneous record of the pulses derived from a separate phase-switching receiver operating with equal cables in the usual fashion. Also shown, in Fig. 1*c*, is a simulated record obtained with exactly the same arrangement but using a signal generator, instead of the source, to provide the swept frequency. For observation with $l > 450$ m the periodic oscillations were slowed down to a low frequency by an additional phase shifting device in order to prevent severe attenuation of the output signal by the time constant of the receiver. The rate of change of signal frequency has been deduced from the additional phase shift required and is $\text{d}\nu/\text{d}t = -4·9 \pm 0·5$ MHz s⁻¹. The direction of the frequency drift can be obtained from the phase of the oscillation on the record and is found to be from high to low frequency in agreement with the first result.

The instantaneous bandwidth of the signal may also be obtained from records of the type shown in Fig. 1*b* because the oscillatory response as a function of delay is a measure of the autocorrelation function, and hence of the Fourier transform, of the power spectrum of the radiation. The results of the measurements are displayed in Fig. 3 from which the instantaneous bandwidth of the signal to exp (− 1), assuming a Gaussian energy spectrum, is estimated to be 80 ± 20 kHz.

Pulse Recurrence Frequency and Doppler Shift

By displaying the pulses and time pips from *MSF* Rugby on the same record the leading edge of a pulse of

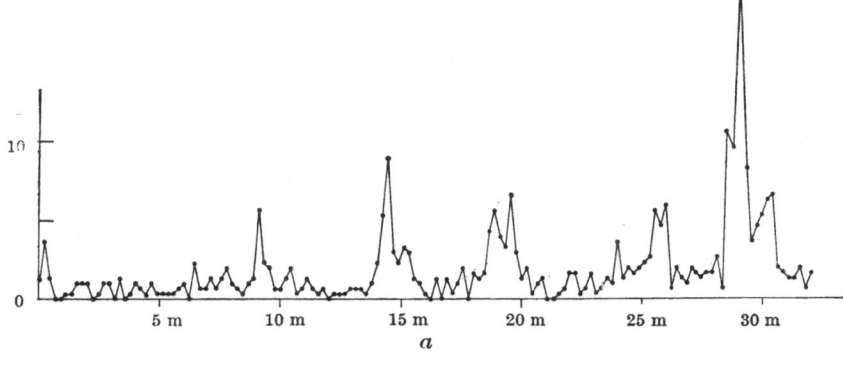

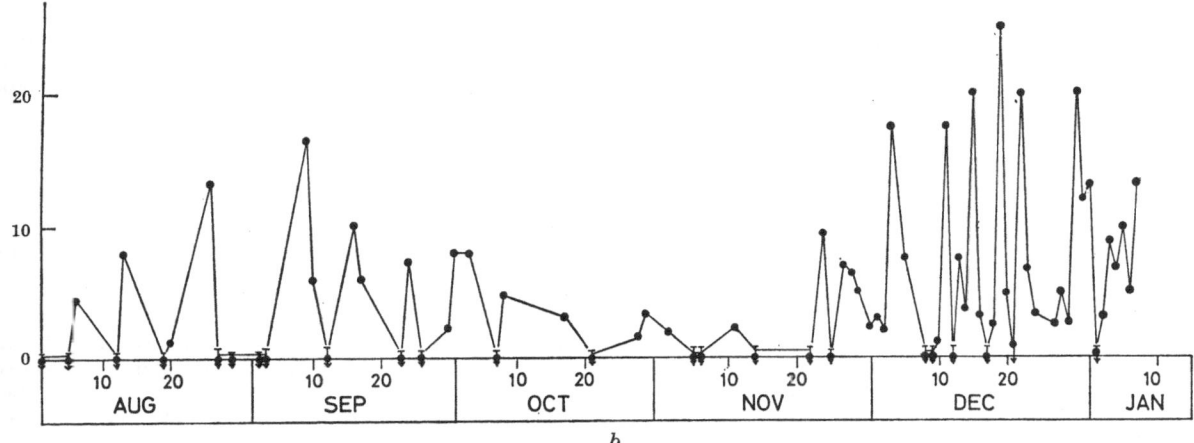

Fig. 2. *a*, The time variation of the smoothed (over ten pulses) pulse amplitude. *b*, Daily variation of peak pulse amplitude. (Ordinates are in units of $W\ m^{-2}\ Hz^{-1} \times 10^{-26}$.)

reasonable size may be timed to an accuracy of about 0·1 s. Observations over a period of 6 h taken with the tracking system mentioned earlier gave the period between pulses as $P_{\mathrm{obs}} = 1\cdot33733 \pm 0\cdot00001$ s. This represents a mean value centred on December 18, 1967, at 14 h 18 m UT. A study of the systematic shift in the frequency of the pulses was obtained from daily measurements of the time interval T between a standard time and the pulse immediately following it as shown in Fig. 4. The standard time was chosen to be 14 h 01 m 00 s UT on December 11 (corresponding to the centre of the reception pattern)

and subsequent standard times were at intervals of 23 h 56 m 04 s (approximately one sidereal day). A plot of the variation of T from day to day is shown in Fig. 4. A constant pulse recurrence frequency would show a linear increase or decrease in T if care was taken to add or subtract one period where necessary. The observations, however, show a marked curvature in the sense of a steadily increasing frequency. If we assume a Doppler shift due to the Earth alone, then the number of pulses received per day is given by

$$N = N_0 \left(1 + \frac{v}{c} \cos \varphi \sin \frac{2\pi n}{366\cdot25} \right)$$

where N_0 is the number of pulses emitted per day at the source, v the orbital velocity of the Earth, φ the ecliptic latitude of the source and n an arbitrary day number obtained by putting $n = 0$ on January 17, 1968, when the Earth has zero velocity along the line of sight to the source. This relation is approximate since it assumes a circular orbit for the Earth and the origin $n = 0$ is not exact, but it serves to show that the increase of N observed can be explained by the Earth's motion alone within the accuracy currently attainable. For this purpose it is convenient to estimate the values of n for which $\delta T / \delta n = 0$, corresponding to an exactly integral value of N. These occur at $n_1 = 15\cdot8 \pm 0\cdot1$ and $n_2 = 28\cdot7 \pm 0\cdot1$, and since N is increased by exactly one pulse between these dates we have

$$1 = \frac{N_0 v}{c} \cos \varphi \left[\sin \frac{2\pi n_2}{366\cdot25} - \sin \frac{2\pi n_1}{366\cdot25} \right]$$

This yields $\varphi = 43° \ 36' \pm 30'$ which corresponds to a declination of $21° \ 58' \pm 30'$, a value consistent with the

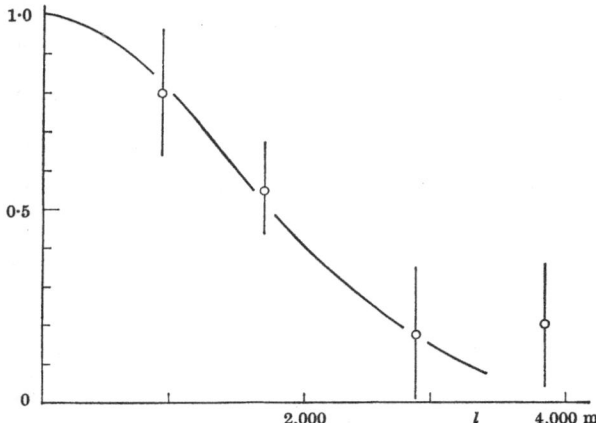

Fig. 3. The response as a function of added path in one side of the interferometer.

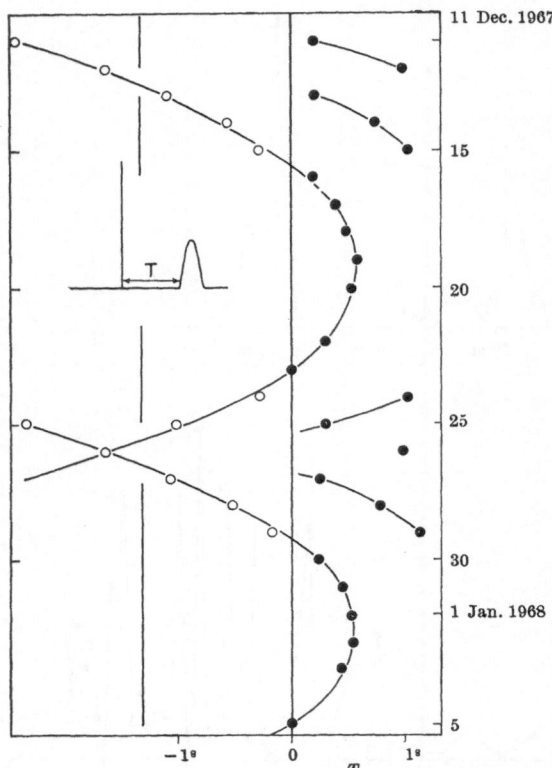

Fig. 4. The day to day variation of pulse arrival time.

declination obtained directly. The true periodicity of the source, making allowance for the Doppler shift and using the integral condition to refine the calculation, is then

$$P_0 = 1 \cdot 3372795 \pm 0 \cdot 0000020 \text{ s}$$

By continuing observations of the time of occurrence of the pulses for a year it should be possible to establish the constancy of N_0 to about 1 part in 3×10^8. If N_0 is indeed constant, then the declination of the source may be estimated to an accuracy of $\pm 1'$; this result will not be affected by ionospheric refraction.

It is also interesting to note the possibility of detecting a variable Doppler shift caused by the motion of the source itself. Such an effect might arise if the source formed one component of a binary system, or if the signals were associated with a planet in orbit about some parent star. For the present, the systematic increase of N is regular to about 1 part in 2×10^7 so that there is no evidence for an additional orbital motion comparable with that of the Earth.

The Nature of the Radio Source

The lack of any parallax greater than about $2'$ places the source at a distance exceeding 10^3 A.U. The energy emitted by the source during a single pulse, integrated over 1 MHz at 81·5 MHz, therefore reaches a value which must exceed 10^{17} erg if the source radiates isotropically. It is also possible to derive an upper limit to the physical dimension of the source. The small instantaneous bandwidth of the signal (80 kHz) and the rate of sweep ($-4 \cdot 9$ MHz s^{-1}) show that the duration of the emission at any given frequency does not exceed $0 \cdot 016$ s. The source size therefore cannot exceed $4 \cdot 8 \times 10^3$ km.

An upper limit to the distance of the source may be derived from the observed rate of frequency sweep since impulsive radiation, whatever its origin, will be dispersed during its passage through the ionized hydrogen in interstellar space. For a uniform plasma the frequency drift

caused by dispersion is given by

$$\frac{d\nu}{dt} = -\frac{c}{L}\frac{\nu^3}{\nu_p{}^2}$$

where L is the path and ν_p the plasma frequency. Assuming a mean density of $0 \cdot 2$ electron cm^{-3} the observed frequency drift ($-4 \cdot 9$ MHz s^{-1}) corresponds to $L \sim 65$ parsec. Some frequency dispersion may, of course, arise in the source itself; in this case the dispersion in the interstellar medium must be smaller so that the value of L is an upper limit. While the interstellar electron density in the vicinity of the Sun is not well known, this result is important in showing that the pulsating radio sources so far detected must be local objects on a galactic distance scale.

The positional accuracy so far obtained does not permit any serious attempt at optical identification. The search area, which lies close to the galactic plane, includes two twelfth magnitude stars and a large number of weaker objects. In the absence of further data, only the most tentative suggestion to account for these remarkable sources can be made.

The most significant feature to be accounted for is the extreme regularity of the pulses. This suggests an origin in terms of the pulsation of an entire star, rather than some more localized disturbance in a stellar atmosphere. In this connexion it is interesting to note that it has already been suggested[2,3] that the radial pulsation of neutron stars may play an important part in the history of supernovae and supernova remnants.

A discussion of the normal modes of radial pulsation of compact stars has recently been given by Meltzer and Thorne[4], who calculated the periods for stars with central densities in the range 10^5 to 10^{19} g cm^{-3}. Fig. 4 of their paper indicates two possibilities which might account for the observed periods of the order 1 s. At a density of 10^7 g cm^{-3}, corresponding to a white dwarf star, the fundamental mode reaches a minimum period of about 8 s; at a slightly higher density the period increases again as the system tends towards gravitational collapse to a neutron star. While the fundamental period is not small enough to account for the observations the higher order modes have periods of the correct order of magnitude. If this model is adopted it is difficult to understand why the fundamental period is not dominant; such a period would have readily been detected in the present observations and its absence cannot be ascribed to observational effects. The alternative possibility occurs at a density of 10^{13} g cm^{-3}, corresponding to a neutron star; at this density the fundamental has a period of about 1 s, while for densities in excess of 10^{13} g cm^{-3} the period rapidly decreases to about 10^{-3} s.

If the radiation is to be associated with the radial pulsation of a white dwarf or neutron star there seem to be several mechanisms which could account for the radio emission. It has been suggested that radial pulsation would generate hydromagnetic shock fronts at the stellar surface which might be accompanied by bursts of X-rays and energetic electrons[2,3]. The radiation might then be likened to radio bursts from a solar flare occurring over the entire star during each cycle of the oscillation. Such a model would be in fair agreement with the upper limit of $\sim 5 \times 10^3$ km for the dimension of the source, which compares with the mean value of 9×10^3 km quoted for white dwarf stars by Greenstein[5]. The energy requirement for this model may be roughly estimated by noting that the total energy emitted in a 1 MHz band by a type III solar burst would produce a radio flux of the right order if the source were at a distance of $\sim 10^3$ A.U. If it is assumed that the radio energy may be related to the total flare energy ($\sim 10^{32}$ erg)[6] in the same manner as for a solar flare and supposing that each pulse corresponds to one flare, the required energy would be $\sim 10^{39}$ erg yr^{-1}; at a distance of 65 pc the corresponding value would be $\sim 10^{47}$ erg yr^{-1}. It has been estimated that a neutron star may contain $\sim 10^{51}$ erg in vibrational modes so the energy requirement does not appear unreasonable, although other damping mechanisms are likely to be important when considering the lifetime of the source[4].

The swept frequency characteristic of the radiation is reminiscent of type II and type III solar bursts, but it seems unlikely that it is caused in the same way. For a white dwarf or neutron star the scale height of any atmosphere is small and a travelling disturbance would be expected to produce a much faster frequency drift than is actually observed. As has been mentioned, a more likely possibility is that the impulsive radiation suffers dispersion during its passage through the interstellar medium.

More observational evidence is clearly needed in order to gain a better understanding of this strange new class of radio source. If the suggested origin of the radiation is confirmed further study may be expected to throw valuable light on the behaviour of compact stars and also on the properties of matter at high density.

We thank Professor Sir Martin Ryle, Dr J. E. Baldwin, Dr P. A. G. Scheuer and Dr J. R. Shakeshaft for helpful discussions and the Science Research Council who financed this work. One of us (S. J. B.) thanks the Ministry of Education of Northern Ireland and another (R. A. C.) the SRC for a maintenance award; J. D. H. P. thanks ICI for a research fellowship.

Received February 9, 1968.

[1] Hewish. A., Scott, P. F., and Wills, D., *Nature*, 203, 1214 (1964).
[2] Cameron, A. G. W., *Nature*, 205, 787 (1965).
[3] Finzi, A., *Phys. Rev. Lett.*, 15, 599 (1965).
[4] Meltzer, D. W., and Thorne, K. S., *Ap. J.*, 145, 514 (1966).
[5] Greenstein, J. L., in *Handbuch der Physik*, L., 161 (1958).
[6] Fichtel, C. E., and McDonald, F. B., in *Annual Review of Astronomy and Astrophysics*, 5, 351 (1967).

2. Observations of some further Pulsed Radio Sources

by
J. D. H. PILKINGTON
A. HEWISH
S. J. BELL
T. W. COLE

Mullard Radio Astronomy Observatory,
Cavendish Laboratory,
University of Cambridge

In a recent communication[1] an account was given of the discovery of a new class of radio source characterized by the emission of short pulses of radiation having an extremely constant repetition frequency. The records on which the source was first detected were taken during a survey for the investigation of compact radio sources using the method of interplanetary scintillation. Following the recognition of the first pulsed source the survey records, which covered the region $-08° < \delta < 44°$, were examined for evidence of further similar sources. Where these records indicated that the intensity fluctuations of a particular source were more impulsive than those caused by interplanetary scintillation, further observations were made. These led to the discovery of three additional pulsed sources. Even though each area of sky was observed on about twenty separate occasions during this survey,

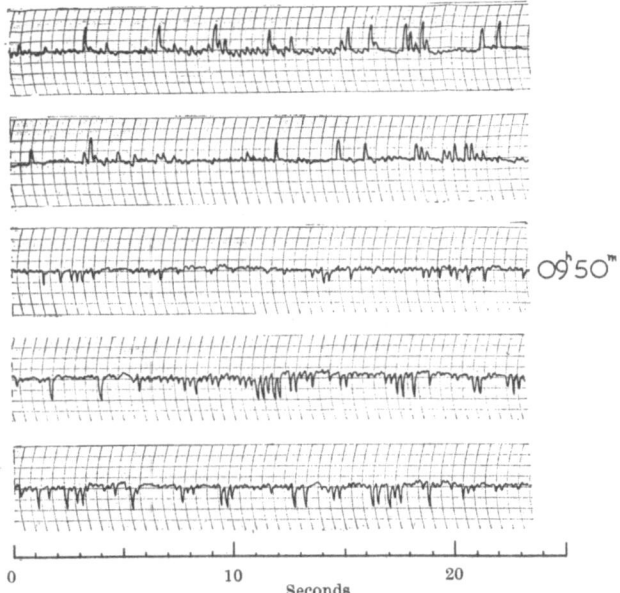

Fig. 2. A continuous set of observations of *CP*.0950 at a time of strong activity, showing the irregular variations of intensity from pulse to pulse. The deflexions change sign as the source moves through the interference pattern of the aerial. Recording time constant 0·1 s.

the large day-to-day variations of flux density from the known sources indicate that this programme should not be regarded as an exhaustive search of the entire region, and observations are continuing.

The three sources and the original one have been given numbers of the form *CP*.1919 to indicate the Cambridge pulsed source at $\alpha = 19^h 19^m$.

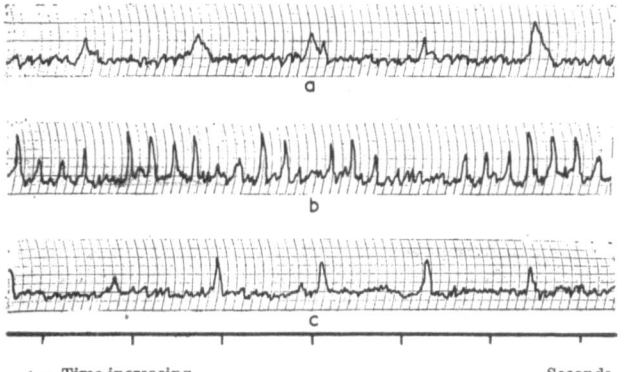

←—Time increasing   Seconds

Fig. 1. Pulses observed with a recording time constant of about 0·03 s on March 21, 1968. (a) *CP*.0834. (b) *CP*.0950, during a period of intense activity. (c) *CP*.1133.

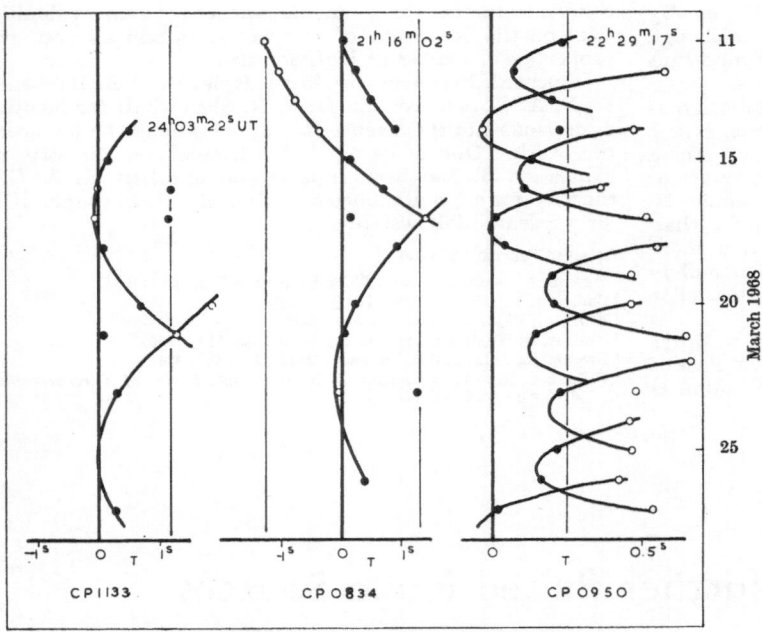

Fig. 3. The day-to-day variation of pulse arrival time.

The three additional sources emit pulses which are remarkably similar to those from the first source, and their characteristics have been obtained by similar methods. Examples of the observed pulses are shown in Figs. 1 and 2. All the measurements were made at a frequency of 81·5 MHz with a bandwidth of 1 MHz, using the 470 m × 45 m north–south phased array and the 470 m × 20 m east–west phased array at the Mullard Radio Astronomy Observatory.

The approximate positions of the sources were derived from the response of the aerial systems. The position of $CP.0950$ has been determined with greater precision from observations at 408 MHz with the one-mile telescope (following communication) using the same method as previously employed[2] for locating $CP.1919$. The best available positions for all four sources are given in Table 1.

The periodicity of the pulses was determined in the same way as previously reported[1]. The time of occurrence of a pulse at approximately the same sidereal time each day was determined, and the incremental time interval between this pulse and a standard time differing by successive units of 23 h 56 m 04 s was plotted as shown in Fig. 3.

In this way the variation of periodicity caused by the motion of the Earth has been compared with the expected variation, and found to be consistent in each case. The true periods P_0 in solar seconds are given in Table 1.

The rate of change of frequency during the pulse and the intrinsic pulse duration have been derived, as before, from observations in which the signal from one aerial of the interferometer was delayed relative to the other by passage through an additional length of cable. In the case of $CP.0950$ it was found that the results were often in-

consistent with the emission of a broad-band pulse; individual pulses sometimes occurred in which nearly all the energy received within a receiver bandwidth of 1 MHz was confined to a band of less than 0·3 MHz. This feature will be discussed in more detail later[3].

The frequency sweep can be interpreted in terms of dispersion in the intervening medium, and the integrated electron density Nl is given in Table 1. The smaller dispersion of $CP.0950$ suggests that for an assumed value of $N \sim 0.1$–0.2 cm^{-3} the distance is likely to be only 15–30 pc.

The short pulse lengths of all the sources indicate physical dimensions in the range 3,000–10,000 km. The variations in the peak pulse amplitude observed from day to day are shown in Fig. 4.

The similarity in the quantities given in Table 1 shows that pulsating sources are, indeed, a new class of object in which the intrinsic powers, pulse widths, variability and periodicities are similar. Although $CP.0950$ has a periodicity somewhat shorter than the other three, and it would be difficult to detect still shorter pulse periods with the present system, there are no observational selection effects which would reduce the probability of detecting sources having periodicities as long as 10 s. The limited range in periodicity clearly has great significance in relation to the nature of the sources, and the shorter period of $CP.0950$ may make it more difficult to account for them without invoking the very high densities of neutron stars.

All the sources are characterized by a period of emission which is very much shorter than the repetition period and which occurs at a precisely defined phase of the cycle. This places limitations on the eccentricity of the orbits in a binary neutron star model[4] because of precession of the perihelion; in the case of $CP.0950$ the eccentricity must be less than about 0·1.

All the sources show an extremely variable flux density both from pulse to pulse and on a longer time scale. In the first communication[1] it was suggested that the rapid variation from pulse to pulse might be caused by interplanetary scintillation. The recent observations were, however, carried out during the night when interplanetary scintillation is known to be small, and it seems that the rapid variations in flux density must be interpreted in terms of the source; the pulse to pulse variation cannot be attributed to interstellar scintillation (unpublished results of P. A. G. Scheuer).

Using the approximate distances given by the measured dispersion, we may conclude that the local density of pulsating sources is $\sim 10^{-5}$ pc^{-3}. The lifetime during which radio pulses are emitted from each source will depend on the supply of energy and on the mechanism by which it is converted into radio pulses; there may be many other "dead" sources which are no longer observable. If the sources are associated with superdense stars of mass $\sim M_\odot$ and if they emit radio pulses for T years, the total number which have occurred over a period of $\sim 10^{10}$ years

	CP.0834	CP.0950	CP.1133	CP.1919
Table 1. CHARACTERISTICS OF THE FOUR PULSED RADIO SOURCES				
$\alpha(1950.0)$	$08^h 34^m 07^s \pm 15^s$	$09^h 50^{\cdot n} 28^s\cdot95 \pm 0^s\cdot7$	$11^h 33^m 32^s \pm 20^s$	$19^h 19^m 37^s\cdot0 \pm 0^s\cdot2$
$\delta(1950.0)$	$07^\circ 00' \pm 45'$	$08^\circ 10' \pm 1'$	$17^\circ 00' \pm 45'$	$21^\circ 47' 02'' \pm 10''$
P_0 (s)	$1\cdot27379 \pm 0\cdot00008$	$0\cdot253071 \pm 0\cdot000008$	$1\cdot1880 \pm 0\cdot0004$	$1\cdot3372795 \pm 0\cdot0000020$
$-\left(\dfrac{d\nu}{dt}\right)$ at 81·5 MHz (MHz s^{-1})	$5\cdot3 \pm 0\cdot5$	20 ± 5	11 ± 3	$5\cdot15 \pm 0\cdot03$ (ref. 6)
Integrated electron density Nl (cm^{-3} pc)	12 ± 1	$3\cdot2 \pm 0\cdot8$	6 ± 2	$12\cdot55 \pm 0\cdot06$ (ref. 6)
Emitted pulse duration (Gaussian) (ms)	35 ± 10	< 10	12 ± 4	16 ± 4
Mean flux density at 81·5 MHz (10^{-26} W m^{-2} Hz^{-1})	0·3	0·8	0·3	0·4
l^{II}	220°	230°	240°	56°
b^{II}	26°	44°	70°	4°

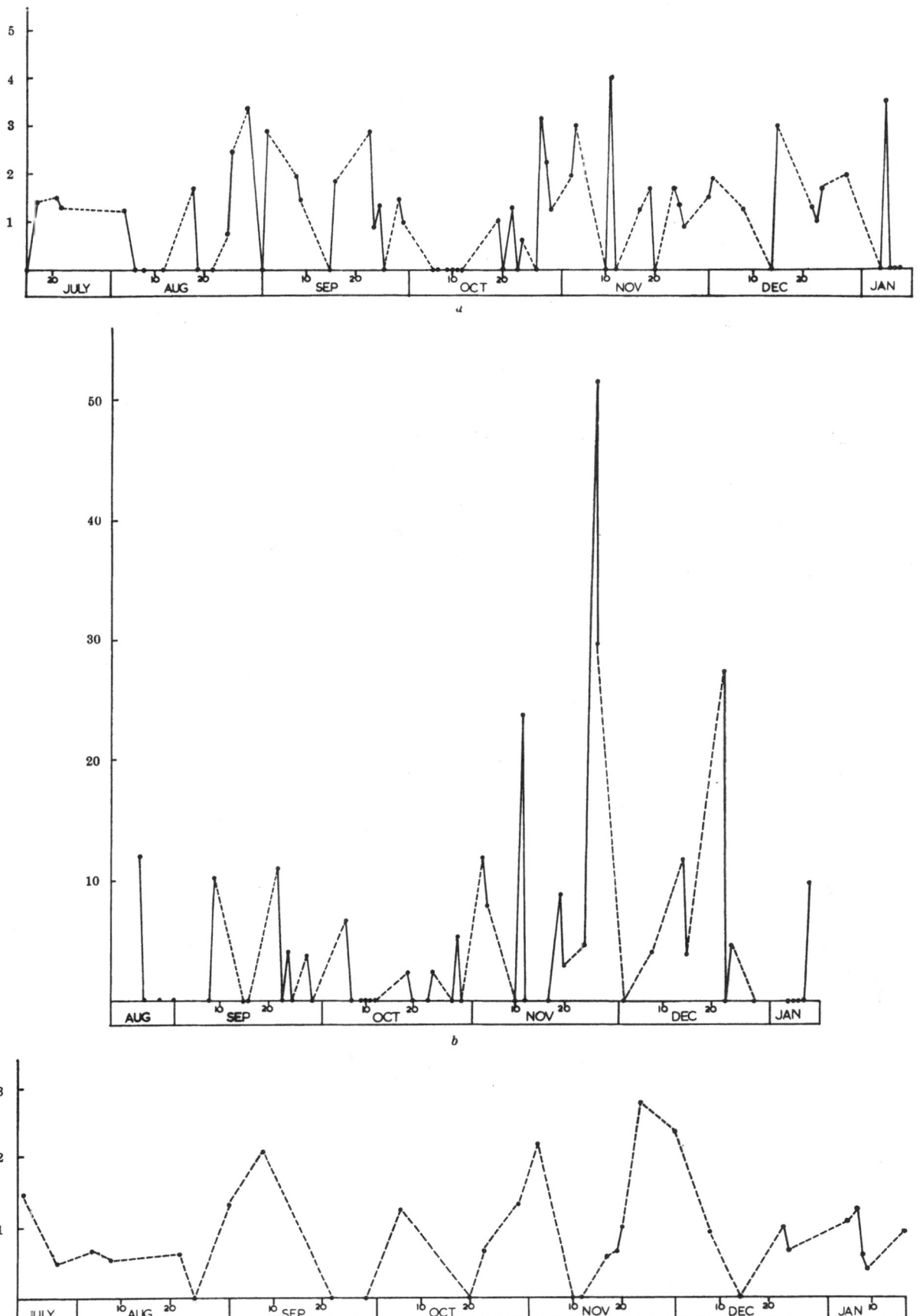

Fig. 4. The day-to-day variation of peak pulse energy. The ordinates are in units of 10^{-26} J m^{-2} Hz^{-1}. (a) CP.0834, (b) CP.0950, (c) CP.1133.

will give a local density of matter of

$$\sim \frac{10^{10}}{T} \cdot 10^{-5} \, M_\odot \, \mathrm{pc}^{-3}$$

Oort[5] has derived an upper limit of 0·14 $M_\odot$ pc⁻³ for all matter in the solar neighbourhood, and we must therefore conclude that the radio pulses must be emitted for a period of at least 6×10^5 years.

If the source of energy is the vibrational modes of a white dwarf or neutron star, a conversion efficiency much higher than those occurring in solar flares must occur. Alternatively some other source model must be found in which a larger supply of energy is available.

We thank Professor Sir Martin Ryle for his encourage-

ment and Dr P. A. G. Scheuer for helpful discussions. One of us (J. D. H. P.) thanks ICI for a research fellowship; S. J. B. and T. W. C. acknowledge maintenance awards from the Northern Ireland Ministry of Education and CSIRO, respectively.

Received April 3, 1968.

[1] Hewish, A., Bell, S. J., Pilkington, J. D. H., Scott, P. F., and Collins, R. A., *Nature*, **217**, 709 (1968), (Paper 1).

[2] Ryle, M., and Bailey, J. A., *Nature*, **217**, 907 (1968).

[3] Scott, P. F., and Collins, R. A., *Nature*, **218**, 230 (1968), (Paper 8).

[4] Saslaw, W. C., Faulkner, J., and Strittmatter, P. A., *Nature*, **217**, 1222 (1968), (Paper 42).

[5] Oort, J. H., *Bull. Astro. Inst. Netherlands*, **15**, 45 (1960).

[6] Davies, J. G., Horton, P. W., Lyne, A. G., Rickett, B. J., and Smith, F. G., *Nature*, **217**, 910 (1968), (Paper 6).

3. Search for Pulsating Radio Sources in the Declination Range $+44° < \delta < +90°$

by
T. W. COLE
J. D. H. PILKINGTON
Mullard Radio Astronomy Observatory,
Cavendish Laboratory,
University of Cambridge

THE survey at 81·5 MHz which led to the discovery of pulsating radio sources at Cambridge[1,2] included declinations in the range $-08° < \delta < 44°$. The antenna will operate at higher declinations, although with reduced sensitivity, and more recent observations have extended the search area to $\delta = +90°$.

The receiver system was the same as that used previously, slowly varying components of the receiver output being removed before recording by a high pass filter cutting off at 0·1 Hz. A receiver time constant of 0·05 s was adopted and the recorder had a time constant of 0·2 s. The receiver bandwidth ($\Delta\nu$) was 1 MHz.

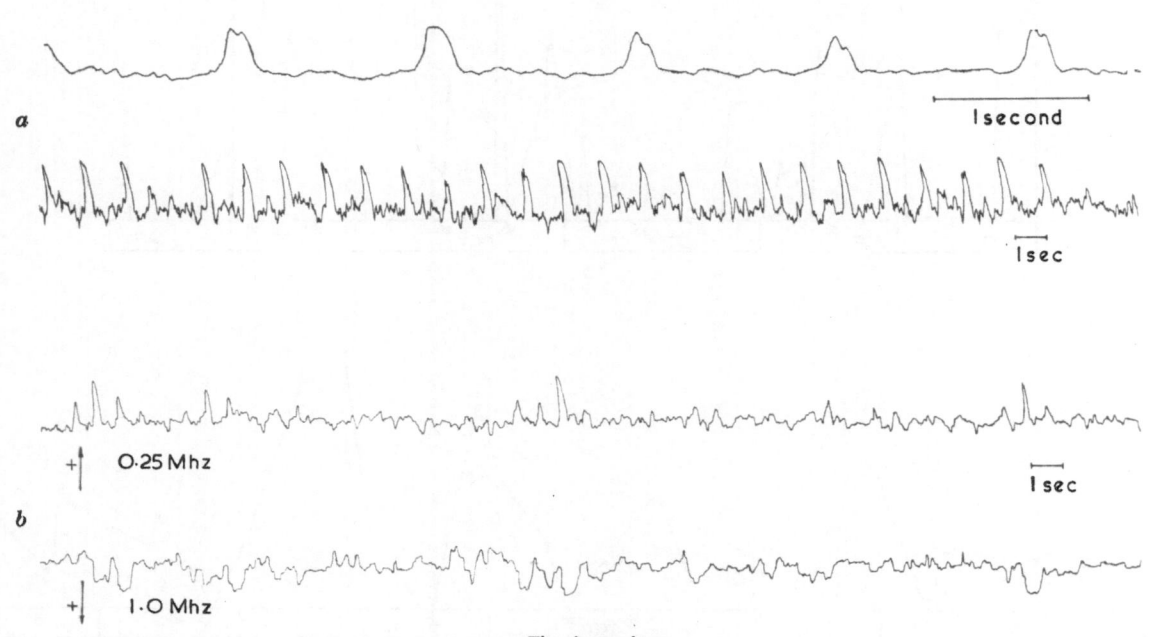

Fig. 1. Series of pulses of (*a*) *CP* 0808 with a 1 MHz bandwidth and (*b*) *CP* 0328 with 1 MHz and 250 kHz bandwidths.

Possible pulsating sources were provisionally distinguished from normal sources exhibiting interplanetary scintillation by visual inspection of the records, and observations of these were made using a fast chart speed and a recording time constant of about 0.06 s; observations were also made with another receiver of bandwidth 0.25 MHz.

The survey may be limited either by the noise level which varies from about 0.5×10^{-26} W m^{-2} Hz^{-1} at $\delta = 44°$ to double this value at $\delta = 90°$, or by too large a dispersion in the received pulses. If dispersion $(d\nu/dt)$ is such that the time for the signal to sweep through the receiver band is greater than the pulse period then the pulses are distinguished only if the pulse to pulse amplitude variation is large; thus sources will only be detected for

$$P \gtrsim \left(\frac{d\nu}{dt} \cdot \frac{1}{\Delta\nu}\right)^{-1}$$

For an assumed interstellar electron density of 0.1 cm^{-3} and a bandwidth of 1 MHz, sources similar to CP 1919 would be detectable within 500 pc while sources like CP 0950 would be detectable within 150 pc.

The complete range $44° < \delta < 90°$ was surveyed on at least six different occasions in order to allow a reasonable chance of detecting sources the intensity of which may vary from day to day. The total observing time for each beam area was thus at least 30 min. Further observations using the high speed recorder were then made of 98 suspected pulsating sources. Of these, two were recognized as genuine and more recent observations have confirmed that another, HP 1506 (ref. 3), is also present.

The positions and radiation characteristics of the new pulsating sources, including an improved value of right ascension and a measurement of dispersion of HP 1506, are shown in Table 1. The source CP 0808 appears to be the strongest source yet found and is similar in most respects to CP 1919. The sources CP 0328 and HP 1506 are of interest because their periods fall midway between that of CP 0950 and the remaining four; if dispersion is a reliable indicator CP 0328 is also the most distant pulsating source yet detected. The mean flux densities quoted are little more than estimates in view of the considerable intensity variations of the sources. The frequency sweep rate was determined by observations with an added delay cable in one interferometer arm; these also revealed frequency structure in both CP 0328 and HP 1506. The pulses from the three sources are illustrated in Figs. 1 and 2. Autocorrelation and pulse height analysis of the source CP 0808 give no evidence for periodicities other than that of the fundamental. The pulse height variation of CP 0328 appears similar to CP 1133.

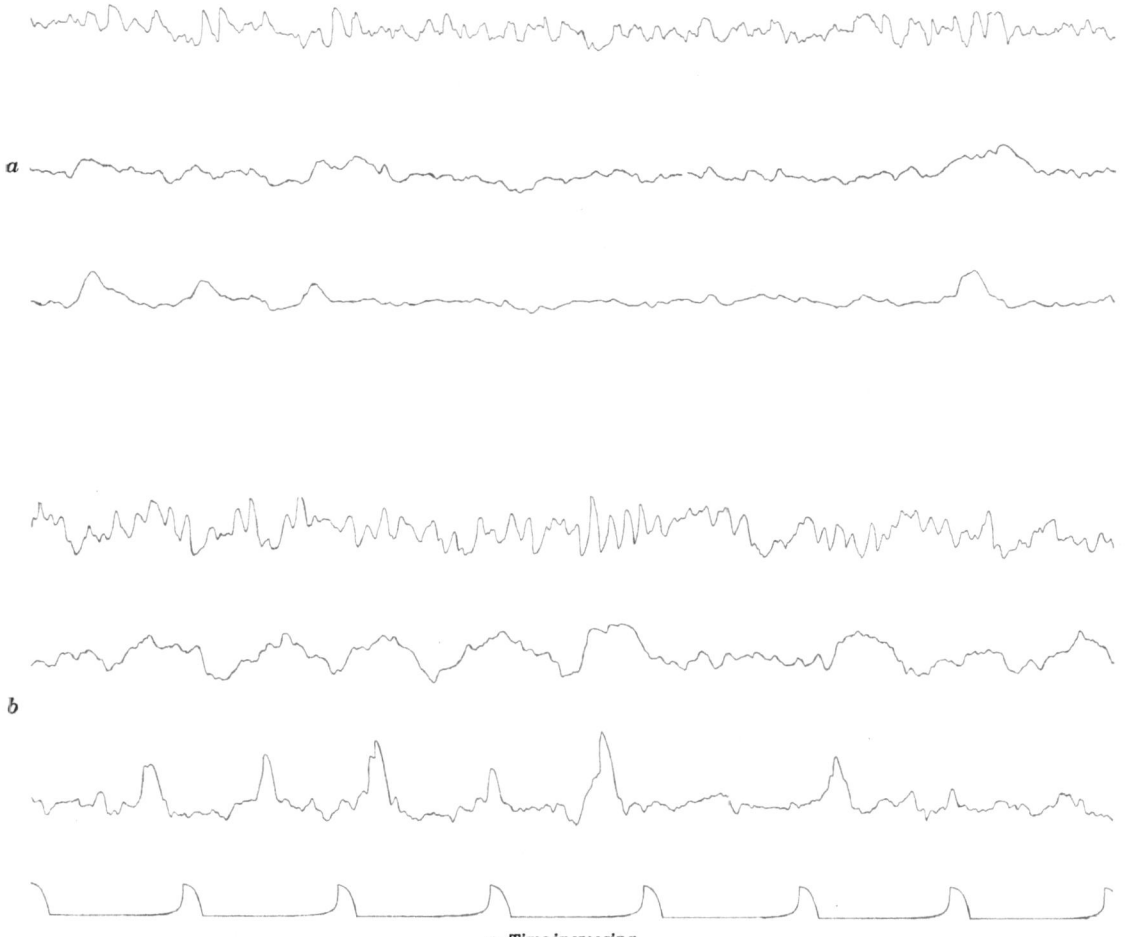

←Time increasing

Fig. 2. (a) Simultaneous records of CP 0328 with a 250 kHz bandwidth on the lower trace; 1.0 MHz on the middle trace; and 1 MHz with 660 metres delay cable on the upper trace. (b) The similar recording of HP 1506 with 1 s markers on the bottom trace.

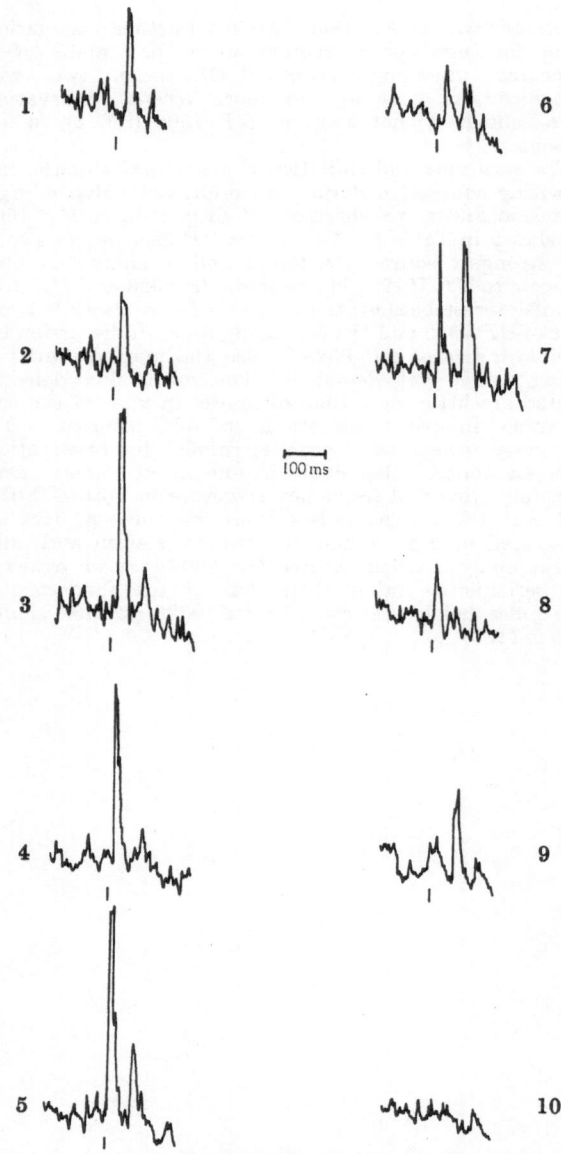

Fig. 3. Ten consecutive pulses of the source CP 0808 with a time resolution of about 8 ms.

Table 1

	CP 0808	CP 0328	HP 1506	
α (1950·0)	08h 08m 50s ± 30s	03h 28m 52s ± 15s	15h 07m 40s ± 30s	
δ (1950·0)	75° 10' ± 30'	55° ± 1°	55½° ± 1°	(3)
l^{II}	140°	145°	90°	
b^{II}	+34°	0°	+53°	
P_0 (s)	1·29223	0·71446	0·7397	(3)
	± 0·00003	± 0·00010	± 0·0001	
$\left(-\dfrac{d\nu}{dt}\right)$ MHz s⁻¹	6·3 ± 0·6	3·6 ± 0·25	4·16 ± 0·25	
Mean flux W m⁻² Hz⁻¹	1·5	0·2	0·3	
Integrated electron density pc cm⁻³	10 ± 1	18 ± 1·5	15·5 ± 1·0	
Distance pc	100	180	150	

Observations of the shape of individual pulses from the source CP 0808 have been made using a total power receiver with a bandwidth of 30 kHz at 81·5 MHz. Dispersion across the band is about 5 ms for this source, and a receiver time constant of 6 ms was used. The receiver output was displayed on a cathode ray oscilloscope, the sweep of which was synchronized to the pulse repetition period, and the trace was photographed on continuously moving film. The pulses are predominantly double with a separation between components of about 60 ms; either component may be present independently of the other and its amplitude may vary widely from pulse to pulse. There is often structure within the components. The total period during which signals may occur is about 100 ms, which is rather longer than the pulse durations previously reported[4] and may indicate that this source is physically larger than the original four. Ten consecutive pulses of CP 0808 observed with this system are shown in Fig. 3.

With a total of seven sources statistical investigations are of low significance but there is as yet no indication that the sources are concentrated towards the galactic equator. This implies that the sources are within the galactic distribution of ionized hydrogen and that the distances derived by assuming $n_e = 0·1$ cm⁻³ are not appreciably underestimated.

We thank Dr A. Hewish for helpful discussions and CSIRO and ICI for support.

Received July 30, 1968.

[1] Hewish, A., Bell, S. J., Pilkington, J. D. H., Scott, P. F., and Collins, R. A., *Nature*, **217**, 709 (1968), (Paper 1).
[2] Pilkington, J. D. H., Hewish, A., Bell, S. J., and Cole, T. W., *Nature*, **218**, 126 (1968), (Paper 2).
[3] *Intern. Astro. Union Circular*, July 1968.
[4] Lyne, A. G., and Rickett, B. J., *Nature*, **218**, 326 (1968), (Paper 10).

4. New Pulsating Radio Source

by

G. R. HUGUENIN   A. HARTAI
J. H. TAYLOR   G. S. F. ORSTEN
L. E. GOAD   A. K. RODMAN

Harvard College Observatory,
Cambridge, Massachusetts

A NEW pulsating radio source has been discovered during the first phase of a systematic search for such objects carried out with the 300 foot transit telescope at the US National Radio Astronomy Observatory (NRAO). The new source lies at the position $\alpha(1950) = 15$h 06m ± 2m, $\delta(1950) = +55° 30' ± 40'$, has a heliocentric pulse-repetition period of 0·73968 ± 0·00002 s, and is similar in many respects to the four pulsars discovered earlier this year by Hewish *et al.*[1] and Pilkington *et al.*[2]. We suggest that the source be called HP 1506, for Harvard pulsating source at right ascension 15h 06m, to conform with the nomenclature used for the Cambridge pulsars.

The survey was carried out at a frequency of 110 MHz during the period June 15 to July 11, 1968. The 300 foot antenna was driven back and forth in declination at a rate of 2·5 arc degrees min⁻¹, so as to trace out a zigzag pattern across the sky. The widths of the declination strips scanned were chosen so that the adjacent end points of the zigzag were separated by approximately one half-power beamwidth, or 2°. Thus, for example, the entire declination strip +45° to +60° could be observed in one day; any given location inside this strip was within the main beam of the antenna for approximately one minute each day.

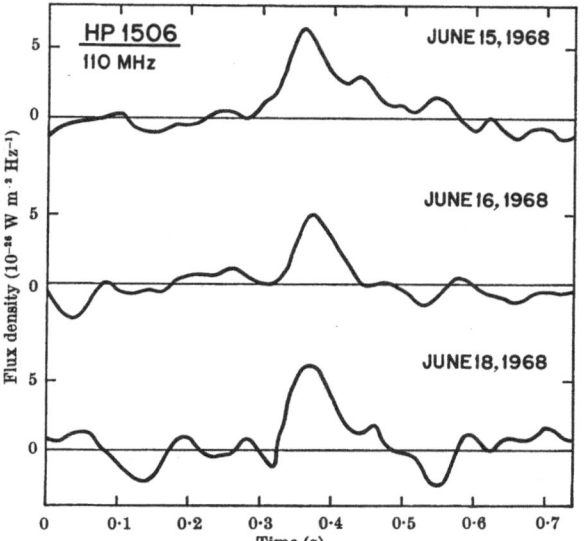

Fig. 1. Average pulses observed from *HP* 1506 on June 15, 16 and 18, 1968. The pulses are broadened by the effects of receiver bandwidth and time constant.

The survey receiver had a bandwidth of 0·3 MHz, and its output was sampled and recorded on magnetic tape once every 10 ms. A search for periodic signals in the receiver output was made with a digital computer. The search programme was most sensitive to periods in the range 0·16 to 2·5 s, although strong signals with periods an octave or so outside this range would also have been detected. Less than 10 per cent of the data taken during the survey have so far been analysed.

Signals from *HP* 1506 were found in the computer analysis of the first day's data, and again in the data recorded on June 16 and 18. Fig. 1 is a plot of the average pulses recorded on these three days. Each curve was obtained by cross-correlating approximately 40 s of data with a train of uniform pulses spaced by 0·7397 s, a figure very close to the correct pulse repetition period. The average pulses shown in Fig. 1 are broadened considerably by the effects of receiver bandwidth and time constant, and do not represent the intrinsic pulse width or shape.

More detailed observations of the source were made later with six receivers operating simultaneously, each having a bandwidth of 0·1 MHz and an effective integration time of 1·6 ms. The six receivers were tuned in pairs to three different frequencies in the range 110 to 116 MHz, and the pairs were connected to orthogonally polarized feed antennae. Preliminary results from these data show (*a*) that the pulses from *HP* 1506 are intrinsically about 20 ms in width; (*b*) that the signals have considerable linear polarization; and (*c*) that the frequency drift rate, presumably due to dispersion in the interstellar medium, is $-8·0\pm0·2$ MHz s⁻¹ at 110 MHz. The latter figure corresponds to an integrated electron density along the path of propagation of $\int N_e\, \mathrm{d}l = 20·0\pm0·5$ cm⁻³ pc. This value is greater than the corresponding values found for the four Cambridge pulsars, and may be indicative of a greater distance.

The intensity of the pulses from *HP* 1506 is comparable with those from *CP* 0834 and *CP* 1133, which are the weaker of the four Cambridge pulsars at frequencies near 100 MHz. The strongest pulses we have recorded from *HP* 1506 have an energy of about 15×10^{-26} J m⁻² Hz⁻¹ at 110 MHz, and an average pulse has an energy smaller by approximately a factor of 30.

After the first observations with the 300 foot transit telescope, an improved period and position were obtained by observing the source for several hours at a time with the NRAO 140 foot telescope. The latter observations were made at frequencies of 111, 234, 256, 405 and 610 MHz during the period July 11 to July 15. The pulse dispersion derived from observations near 110 MHz was substantiated by the observations at higher frequencies.

This work was supported by the US National Science Foundation and by the US National Aeronautics and Space Administration. We are grateful for the use of the facilities at the US National Radio Astronomy Observatory.

Received July 22, 1968.

¹ Hewish, A., Bell, S. J., Pilkington, J. D. H., Scott, P. F., and Collins, R. A., *Nature*, 217, 709 (1968), (Paper 1).
² Pilkington, J. D. H., Hewish, A., Bell, S. J., and Cole, T. W., *Nature*, 218, 126 (1968), (Paper 2).

5. Discovery of Two Southern Pulsars

by
A. J. TURTLE
A. E. VAUGHAN
Cornell-Sydney University Astronomy Centre,
School of Physics,
University of Sydney

Two southern pulsars have been discovered by observations at the Molonglo Radio Observatory. One is within 3° of the direction of the galactic centre. Preliminary values for the positions and pulse periods are given in Table 1.

The sources were detected during a survey designed for this purpose which used the east–west arm of the one mile cross radio telescope operating at a frequency of 408 MHz.

Single scans at transit have been made over most of the region between right ascensions 15h to 03h, declinations −10° to −35°. Pulsars are so variable in their intensity that strong sources could easily have been missed because of their quiescence at transit. The east and west sections of the aerial can be operated independently and each has a beamwidth of 5′ in right ascension and 4° in declination; for this survey they were set one half a beamwidth apart in declination. With the low noise preamplifiers installed recently, the peak to peak noise fluctuations on a recorder with a time constant of ~0·1s are equivalent to a flux density of about 2×10^{-26} W m^{-2} Hz^{-1}.

The time of transit of a pulsating object gives an approximate right ascension and the declination was estimated initially from the ratio of the pulse amplitudes received on the two arms. The accuracy of this position was sufficient to allow the full cross to be directed at the appropriate region on later transits. The final positions were derived from observations in July 1968, with the cross and the fan beams of the north–south arm[1]. The results are summarized in Table 1.

Table 1. CHARACTERISTICS OF THE TWO NEW PULSARS

	PSR 1749–28	PSR 2045–16
Right ascension (1950·0)	17h 49m 48·8s ± 0·3s	20h 45m 47·6s ± 0·4s
Declination (1950·0)	−28° 05′ 57″ ± 8″	−16° 27′ 50″ ± 12″
Pulse period (s)	0·5621	1·961
Galactic longitude l^{II}	1·6°	30·5°
Galactic latitude b^{II}	−1·0°	−33·1°

Many observatories will presumably be finding pulsars, and so we suggest that a uniform system of designation be adopted. We propose the contraction PSR (from pulsar) followed by an abbreviated 1950·0 right ascension and declination; these two sources would then be called PSR 1749–28 and PSR 2045–16.

Although we have at present only limited observations of the two new pulsars, the variations in the amplitudes of the pulses are markedly different. Nor do they closely resemble the characteristics of any of the first four pulsars to be discovered. The one with the fast period, PSR 1749–28, has been detected at every transit although there are still pronounced pulse to pulse variations. PSR 2045–16 is often undetectable, but when it is active it may produce pulses which are more energetic than those observed from any other pulsar at this frequency. The

1·961s period of PSR 2045–16 is the longest yet announced. The periods of these objects are spread randomly through a range of at least 8 to 1 and any theory of the origin of the pulsed emission must be sufficiently noncritical to accommodate this result.

The variation with frequency of the time of arrival of a pulse and its polarization can be used to study the properties of the interstellar plasma[2,3]. It is now possible to use PSR 1749–28 to investigate these properties in a direction towards the galactic centre. In particular, it should be possible to establish unambiguously that the source is within the galaxy. There could be some doubt for the sources at high galactic latitudes because of the uncertainty in the mean electron density.

A search of the Palomar Sky Survey plates has been made and the positions of possible identifications have been measured to an accuracy of a few seconds of arc. In neither case is there a likely identification. The only object within 20″ of the radio position of PSR 2045–16 has a magnitude of 19·5 and lies just outside the error rectangle at 20h 45m 48·2s, −16° 27′ 54″. The other pulsar is situated in a crowded field, but there is only one possible object (18th magnitude) which is just inside the rectangle at 17h 49m 49·0s, −28° 05′ 50″. There are three other objects close to the rectangle including one about 15th magnitude at 17h 49m 48·4s, −28° 05′ 54″.

We thank Professor B. Y. Mills for his advice and cooperation and R. W. Hunstead for assistance with the observations. This work was supported by grants from the Australian Research Grants Committee and the US National Science Foundation. One of us (A. E. V.) holds a Commonwealth postgraduate research studentship.

Received August 6, 1968.

[1] Turtle, A. J., and Vaughan, A. E., Nature, 219, 689 (1968), (Paper 5).
[2] Tanenbaum, B. S., Zeissig, G. A., and Drake, F. D., Science, 160, 760 (1968).
[3] Smith, F. G., Nature, 218, 325 (1968), (Paper 47).

Signal Characteristics

6. Pulsating Radio Source at $\alpha = 19^h 19^m$, $\delta = +22°$

by
J. G. DAVIES
P. W. HORTON
A. G. LYNE
B. J. RICKETT
F. G. SMITH
University of Manchester,
Nuffield Radio Astronomy Laboratories,
Jodrell Bank

AT the time of publication of the remarkable discovery by a group of Cambridge radio astronomers[1] of the pulsating radio source at $\alpha = 19^h 19^m 39^s \pm 3^s$, $\delta = +22° \pm 30′$ (1950), the mark I radio telescope was equipped for reception at frequencies of 151, 240 and 408 MHz. The receivers were intended for studies of flare stars and interplanetary scintillations, and an Argus 400 digital computer was in use for the on-line analysis of the receiver outputs. It was therefore decided to search for radio pulses from this source at these receiver frequencies, which were above that of the original observation at 81 MHz. Radio pulses were

found at all three frequencies, the first observation being on February 24, 1968. The position of the source agreed well with the original observations; the declination was measured as 21° 40′ ± 7′ (1950).

The method of analysis was to obtain an average pulse shape at each frequency by superposing receiver outputs over a number of cycles of the pulse repetition period of 1.33724 s. Each cycle was divided into shorter intervals, originally of 20 ms duration. When the position of the pulse within the period became clear, the interval was reduced progressively and the time scale was expan-

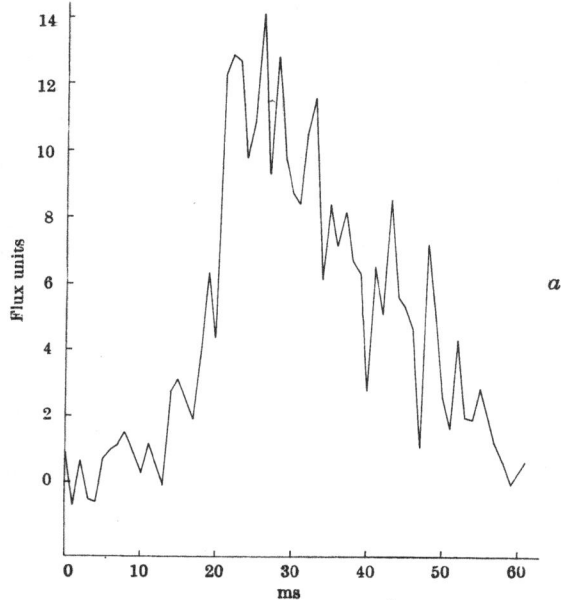

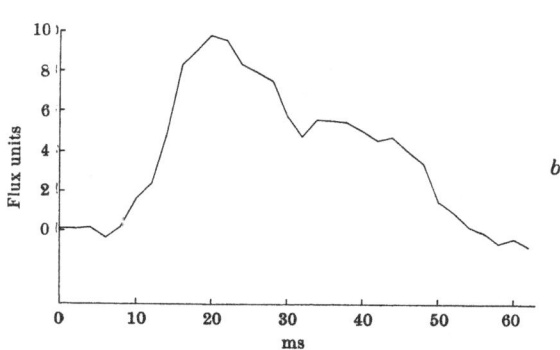

Fig. 1. *a*, The mean pulse shape at 408 MHz on February 28, 1968, at 0455 UT averaged over 2 min. The receiver bandwidth was 1 MHz and the time constant 1 ms. *b*, The mean pulse shape at 408 MHz on February 24, 1968, at 0350 UT averaged over 12 min with the same receiver parameters.

ded, until for some experiments only 64 ms of each period were examined, with a time resolution of 1 ms. Most of the observations were made using three receiver channels simultaneously, recording the average in 32 intervals of 10 ms on each channel. Integration of pulses by this technique usually lasted for periods of 2 min.

Observations were then directed toward measurements of pulse shape at the three frequencies, and to the measurement of the relative times of arrival. The source was also followed for some hours on several days to record the variations of pulse amplitude which occurred on all three frequencies.

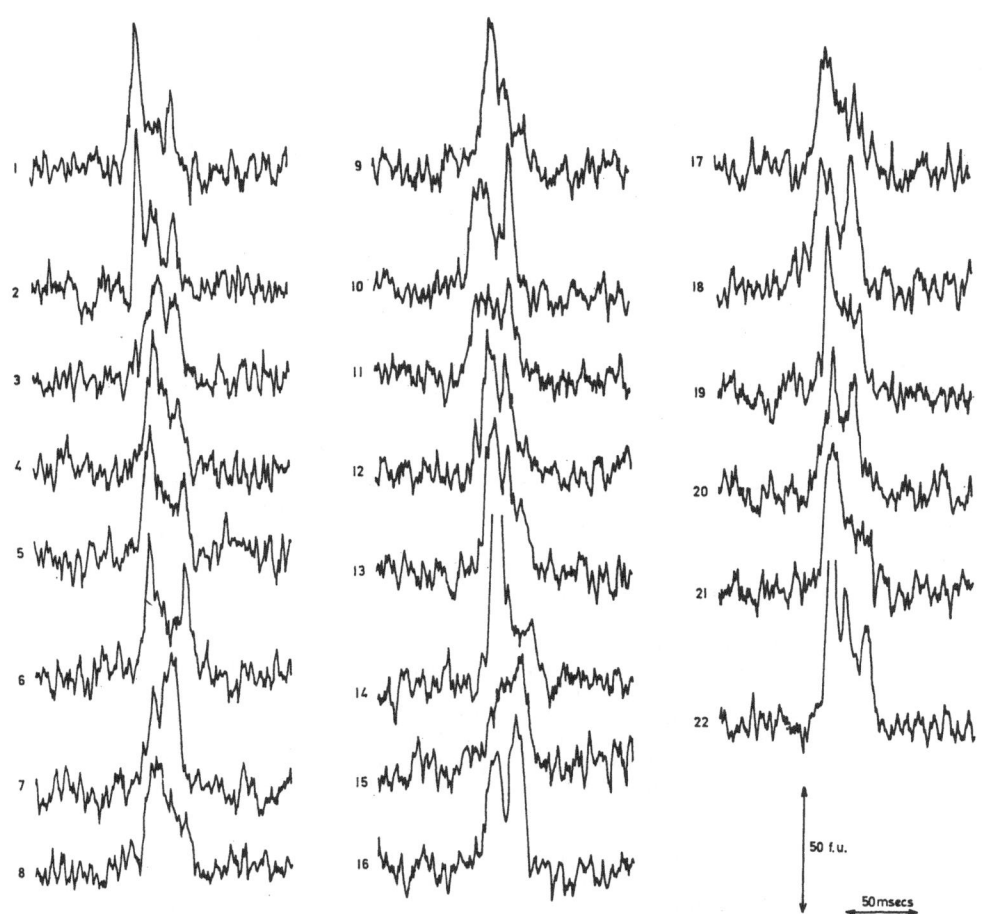

Fig. 2. A series of pulses recorded directly at 408 MHz (bandwidth 4 MHz, time constant 1 ms) on February 29, 1968.

Shape of the Pulses

Fig. 1 shows mean pulse shapes at 408 MHz. Individual pulses at 408 MHz have also been recorded directly on a fast ultraviolet recorder, using a receiver time constant of 1 ms. These are shown in Fig. 2.

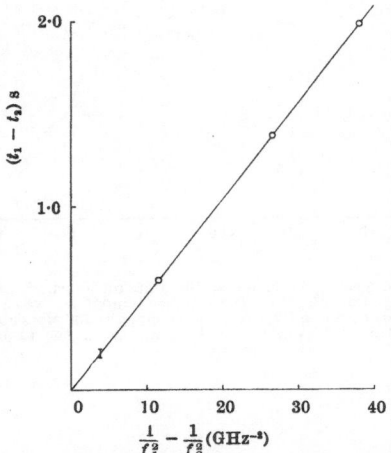

Fig. 3. Dispersion in arrival time of pulses at 151, 240 and 408 MHz. The errors are too small to show. An additional point at 0·2 s delay is obtained from the Cambridge observation at 80·5 and 81·5 MHz.

The time of onset within the period of 1,337 ms is, however, stable within about 5 ms, a remarkably small limit considering the wide variability of pulse amplitude which is observed on all frequencies.

Pulse Delay

Differences in arrival times of the sharp leading edge could be determined with an accuracy of about 5 ms. We therefore tested the suggestion that the dispersion in arrival time on the three frequencies was due to group delay in the propagation path through ionized interstellar gas. If the pulse at frequency f_1 was emitted at time T_1 and observed at time t_1, then

$$t_1 - T_1 = A + \frac{B_2}{f_1^2}$$

where A is the free-space travel time and B is the dispersion constant. Fig. 3 shows the difference $t_1 - t_2$ plotted against $\left(\frac{1}{f_1^2} - \frac{1}{f_2^2}\right)$. The points lie on a straight line whose intercept with the vertical axis is 0 ± 5 ms, showing that the differences in emission times over the frequency band are either very small or that they themselves follow a similar dispersion law. We suggest that the emission in fact takes place nearly simultaneously on all frequencies.

The observed dispersion constant B yields a measure of

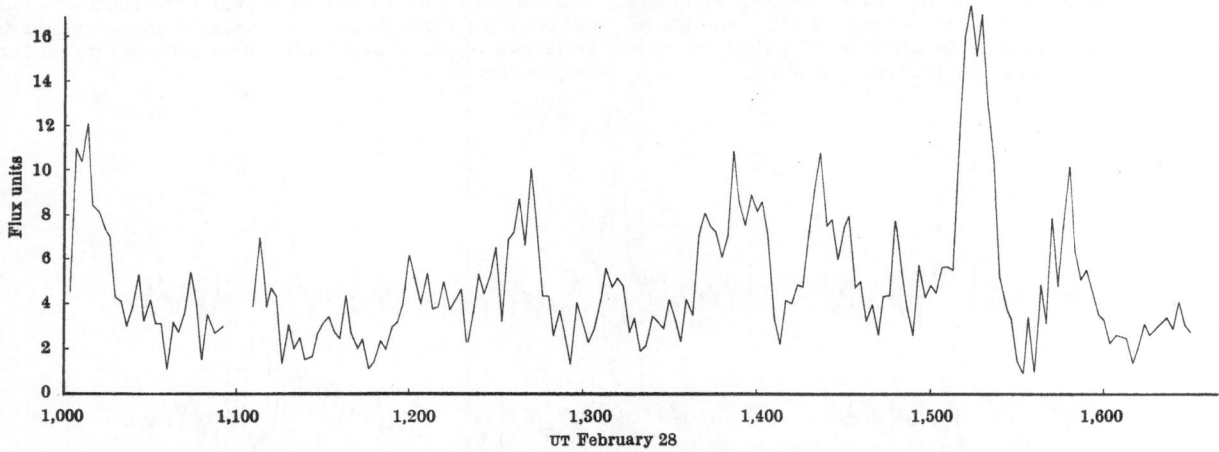

Fig. 4. The variation of pulse amplitude at 408 MHz, averaged over 2 min, for an observing period of 6 h on February 28, 1968.

The average pulse shows a very sharp rise on all frequencies. This steep rise may have been appreciably slowed in these measurements by the differential time of arrival across the receiver bandwidth, which was 1 MHz. At 151 MHz this effect would stretch the pulse by 22 ms, at 240 MHz by 7 ms, and at 408 MHz by 1·5 ms. It appears that the initial rise may in fact occur within 2 ms at 408 MHz, and there is no evidence that the rise is slower at the lower frequencies.

The pulse lasts for about 35 ms on all three frequencies. This duration is consistent with the measurements on 81 MHz, where a delay technique was used to measure the autocorrelation function of the pulse envelope. The autocorrelation function was originally interpreted in terms of a gaussian shaped pulse with width 17 ms. The roughly triangular pulse shape now observed gives a similar width of autocorrelation function, and we conclude that the duration of the pulse is in fact the same at all frequencies so far observed. There is no sign of any impulsive emission outside the normal duration of the pulse.

The measured pulse shape is variable in detail, both between successive pulses, as seen in Fig. 2, and also on averaged pulse shapes. Fluctuations are observed on time scales down to 1 ms, the limit of time resolution.

the integrated electron density in the line of sight equal to $12·55 \pm 0·06$ pc cm^{-3}.

Pulse Amplitude

It is difficult to determine the spectrum of the radio emission, because the amplitudes on all three frequencies are variable, and the variations are not obviously related between the three frequencies. On all three frequencies the flux density during the pulse was usually between 2 and 20×10^{-26} W m^{-2} Hz^{-1}, measured as the peak of the mean profile averaged over 2 min. The direct recordings of single pulses on 408 MHz show short peaks of over 100×10^{-26} W m^{-2} Hz^{-1}.

Fig. 4 shows the variation of pulse amplitude on 408 MHz for a period of 6 h on February 28, 1968. Variations occur on a wide range of time scales. The most prominent feature is a rapid rise by a factor of more than 10, which may persist for 10 min or more and recur after 1–2 h. Variations over a period of 10 s seem to be less than those reported at 81 MHz. No truly periodic variability has been observed.

A receiver at 2,695 MHz on the mark II telescope was also used, and it soon appeared that the flux density of the pulses was much lower at this high frequency. Observations for a period of 3·25 h (1316–1631 UT on February 29, 1968) would have detected a flux density during the pulse, averaged over any one minute, of 0·1 × 10⁻²⁶ W m⁻² Hz⁻¹, but nothing was detected on this frequency.

Nature of the Source

The observation of the nearly simultaneous onset of emission of the pulses on three widely spaced frequencies suggests that the pulses on the various frequencies originate in regions which are separated by less than 1,000 km. Again, the total duration of the pulse, which is the same at all four frequencies, indicates that all sources of the emission lie within a distance of 10,000 km. These distances are consistent with an origin in the very condensed corona of a collapsed star, such as a white dwarf or neutron star, as suggested in ref. 1.

If the origin of the radiation is at a distance of 60 pc, as was suggested to explain the dispersion of arrival time, the simultaneous generation of pulses over a wide bandwidth implies a radiated pulse power of about 10²¹ W and a pulse energy of 2 × 10¹⁹ J. If the source is a star with radius 10,000 km, the power flux at the surface is about 3 × 10⁵ W m⁻², corresponding to a field strength of 10⁴ V m⁻¹. These values would, of course, be even larger for a smaller star.

Received March 2, 1968.

¹ Hewish, A., Bell, S. J., Pilkington, J. D. H., Scott, P. F., and Collins, R. A., *Nature*, 217, 709 (1968), (Paper 1).

7. Linear Polarization in Pulsating Radio Sources

by
A. G. LYNE
F. G. SMITH

University of Manchester,
Nuffield Radio Astronomy Laboratories,
Jodrell Bank

THE mechanism of emission from the pulsating radio sources reported by Hewish *et al.*[1] seems to be quite different from that of any other celestial radio source, particularly because of the wide frequency band over which intense radiation is emitted simultaneously[2]. An analogy may be found, however, in the radio pulse emitted by a cosmic ray shower[3,4], in which a sheet of particles moving relativistically may emit linearly polarized radiation when they are deflected by the Earth's magnetic field. It seemed appropriate therefore to search for linear polarization in the pulsating stars, and we here report the successful detection of linear polarization in all of the four known sources.

Initial tests on the first source CP 1919 at α = 19ʰ 19ᵐ δ = 21° 47′ were disappointing. Two orthogonal dipoles in the Mark I radio telescope were connected to separate receivers at 408 MHz, and the strength of the recorded pulse averaged over 1 min was compared in the two receiver channels. No significant difference was found, although observations were made at several different position angles. It was realized, however, that the variable amplitude of the pulse might be confusing the observations, and that the complex structure of the pulse might imply different origins for different parts of the pulses so that polarization would only be observed during the rare occasions when individual pulses are strong enough for the structure to be resolved in some detail.

The pulses from the source, at α = 09ʰ 50ᵐ, δ = 8° 4′, are usually stronger and simpler, and seem to have fewer separate components. A recording of individual pulses from this source immediately showed a very high degree of linear polarization. Fast galvanometer recordings of the two receiver outputs at 408 MHz showed a ratio which changed substantially over a period of 5 min, and which also showed changes from pulse to pulse. The ratio is occasionally so high that the polarization must be almost completely linear. The observations were repeated at 151 MHz, and again the pulses appeared to be almost completely linearly polarized, because the ratio of intensities in the two receiver channels changed from 4 : 1 to 0·2 : 1 over a period of 20 min (Fig. 1). This rotation of the angle of polarization is consistent with Faraday rotation in the ionosphere.

Further observations at 408 MHz showed also that the pulses from the source at α = 11ʰ 33ᵐ, δ = 16° 09′ are also highly linearly polarized. The pulses from this source often show several discrete components, and on one occasion when the pulse was strong these separate components were visible on fast galvanometer recordings of both receiver outputs, as shown in Fig. 2. It is evident that all the components are highly linearly polarized, but that the angle of polarization is different for the different components.

There is evidence that the polarization angles of these individual features in the pulses remain nearly constant over periods of minutes.

The existence of several components in a single pulse, each polarized at a different angle, may account for the lack of polarization in the source CP 1919 in our initial observations. A more careful examination, using the

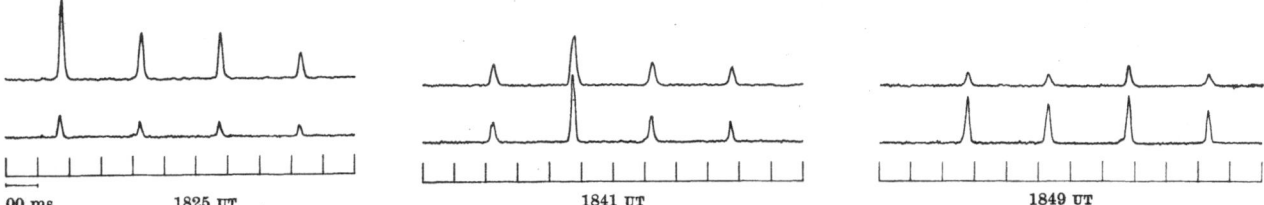

Fig. 1. Fast galvanometer recordings of the outputs of two receivers at 151 MHz connected to orthogonal dipoles at the focus of the Mark I telescope. These were made on the source at α = 09ʰ 50ᵐ, δ = 8° 15′ on April 1, 1968. The change of relative amplitudes in the two channels indicates a linear polarization in these pulses.

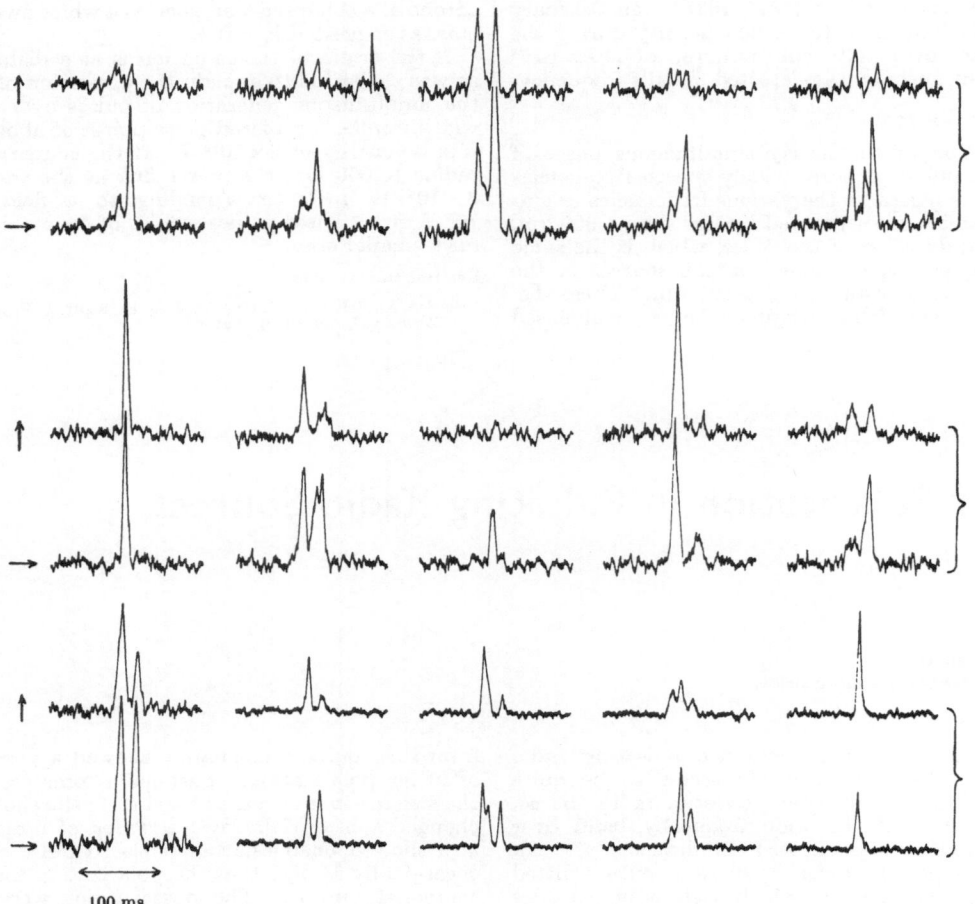

Fig. 2. Fast galvanometer recordings of the outputs of two receivers at 408 MHz connected to orthogonal dipoles at the focus of the Mark I telescope. These were made on the source at $\alpha = 11^h 33^m$, $\delta = 16° 09'$ on April 3, 1968.

same technique in which pulses were averaged over 1 min, showed a ratio of average intensities in the two channels which changed from 2 : 1 to 0·5 : 1 over a period of 1 h, indicating at least a 50 per cent polarization averaged over the whole pulse. A similar result was obtained for the source at $\alpha = 08^h 34^m$, $\delta = 6° 18'$. We conclude that in these sources the radiation is also linearly polarized, possibly again to a high degree in individual components.

Possible theories of the pulse emission include a sequential excitation of radiation from regions in which the frequency is determined by local conditions, such as the plasma density or the gyrofrequency. It has also been suggested that the radiation is actually continuous, and that the pulses are caused by a gravitational focusing in a binary neutron star[5]. Both these theories should be tested by more detailed investigations of the shape of the pulses, but it seems that neither can account satisfactorily for the existence of linear polarization.

It is clear, on any theory, that the radiation must be from a coherent mechanism. The observed brightness temperature T_b during the pulse of CP 1919 is in excess of 10^{21} °K at 151 MHz if we assume that the distance of the star is 60 pc and that the diameter of the emitting region is 10^4 km, the diameter of a white dwarf. A brightness temperature of 10^{21} °K cannot be the result of incoherent radiation because the corresponding energy of an electron is at least 10^{17} eV. Synchrotron radiation from such high energy electrons in a magnetic field of only 1 gauss would extend up to gamma-ray frequencies rather than show the observed cut-off in the region of 1,000 MHz. On the other hand, incoherent radiation from electrons with lower energies would suffer self-absorption, and could not reach such high brightness temperatures.

We therefore suggest that the radiation originates in a region with at least one dimension smaller than 10 cm, to allow for coherence at frequencies up to 1,000 MHz. This region may travel towards the observer with relativistic velocities. Transfer of the energy from a mechanical impulse to electromagnetic waves seems necessarily to involve a magnetic field, and the radiation will therefore be linearly polarized. Because there is frequently observed to be a high degree of polarization in a series of separate components, each with its own position angle, the radiation at any one time must be from a limited region, which may be replaced by another a few seconds or minutes later.

The duration of the whole pulse then seems to be determined mainly by the dimensions of the source. Individual components are separated by up to 30 ms in the pulses from all four sources, suggesting that the radius of the star cannot be much less than 9,000 km. If this interpretation is correct, coherent radiation is being received from near the limb of the star, at a wide angle to the normal. A coherent mechanism is necessarily directive, and a detailed theory of emission must then allow for emission in a suitable direction.

Finally, we note that the discovery of linear polarization opens up the possibility of measuring the Faraday rotation in the interstellar medium, which can then be combined with the measure of the electron content already available from the frequency dispersion of the arrival time of the pulses, to give a very direct measure of the interstellar magnetic field. Preliminary results of this measurement will be reported in a separate communication.

We thank Dr A. Hewish for advance information on the positions of three pulsating radio stars; we have also been helped by discussions on radiation mechanisms with Professor F. D. Kahn.

Received April 5, 1968.

[1] Hewish, A., Bell, S. J., Pilkington, J. D. H., Scott, P. F., and Collins, R. A., *Nature*, **217**, 709 (1968), (Paper 1).
[2] Davies, J. G., Horton, P. W., Lyne, A. G., Rickett, B. J., and Smith, F. G., *Nature*, **217**, 910 (1968), (Paper 6).
[3] Jelley, J. V., Fruin, J. H., Porter, N. A., Weekes, T. C., Smith, F. G., and Porter, R. A., *Nature*, **205**, 327 (1965).
[4] Kahn, F. D., and Lerche, I., *Proc. Roy. Soc.*, A**289**, 206 (1966).
[5] Saslaw, W. C., Faulkner, J., and Strittmatter, P. A., *Nature*, **217**, 1222 (1968), (Paper 42).

8. Further Observations of Pulsating Radio Sources

by

P. F. SCOTT
R. A. COLLINS

Mullard Radio Astronomy Observatory,
Cavendish Laboratory,
University of Cambridge

FOLLOWING the first announcement of the discovery of a new type of pulsating radio source[1], details of three further objects of this type have been published[2], one of these, *CP* 0950, having an appreciably shorter period than the others. In the original article[1] Hewish *et al.* suggested that the radio emission from these sources might arise

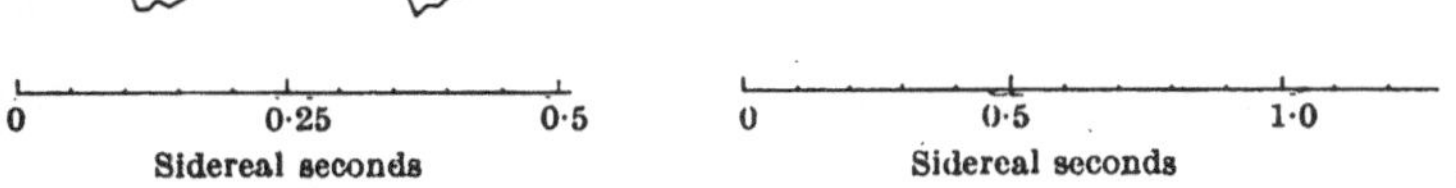

Fig. 1. Combined records of successive pulses for the sources *CP* 0834 and *CP* 0950.

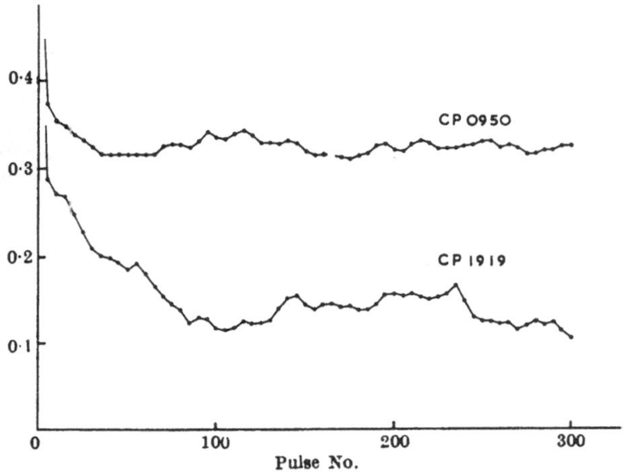

Fig. 2. Autocorrelation functions of pulse amplitude for the sources *CP* 0950 and *CP* 1919.

from the oscillations of a neutron star. More recently, alternative explanations in terms of a binary neutron star system[3] and also a rapidly rotating white dwarf[4] have been proposed. The present communication describes further observations to investigate these possibilities and also to examine the frequency structure of individual pulses; further differences between *CP* 0950 and the other sources have been found.

The observations have been made using the fixed element of the original 178 MHz interferometer at the Mullard Radio Astronomy Observatory[5]. The aerial, 440 m by 20 m and aligned on an east–west axis, has been converted to operate at 81·5 MHz and now incorporates a phasing system which allows the aerial beam to be automatically scanned, at sidereal rate, through an angle of about 10° in right ascension. The first set of observations was made using a simple receiver of 1 MHz bandwidth and centre frequency, 81·5 MHz. The receiver output was recorded on an electroencephalograph recorder with a response extending to 100 Hz, and also digitally on punched paper tape, the signal being first integrated with a time-constant of 0·02 s and then sampled 10 times per sidereal second.

Comparison of Pulse Amplitude Variations with the Binary and Rotating Star Models

The records were first analysed for any systematic difference in the characteristics of odd and even pulses such as might occur if both components of a binary system were radiating. Using the accurately known periods of the pulses, the digitized records of two of the sources (*CP* 0834 and *CP* 0950) were divided into sections of exactly two periods and the sections added together to provide an integrated record of two successive pulses. Each record was analysed for a number of separate periods, each lasting 5 min; the results for two such periods are shown in Fig. 1. None of the results reveals any systematic difference between successive pulses; in the case of *CP* 0950 the average amplitudes agreed within 4 per cent and the time between the two pulses was within 0·2 per cent of the mean period. Comparable

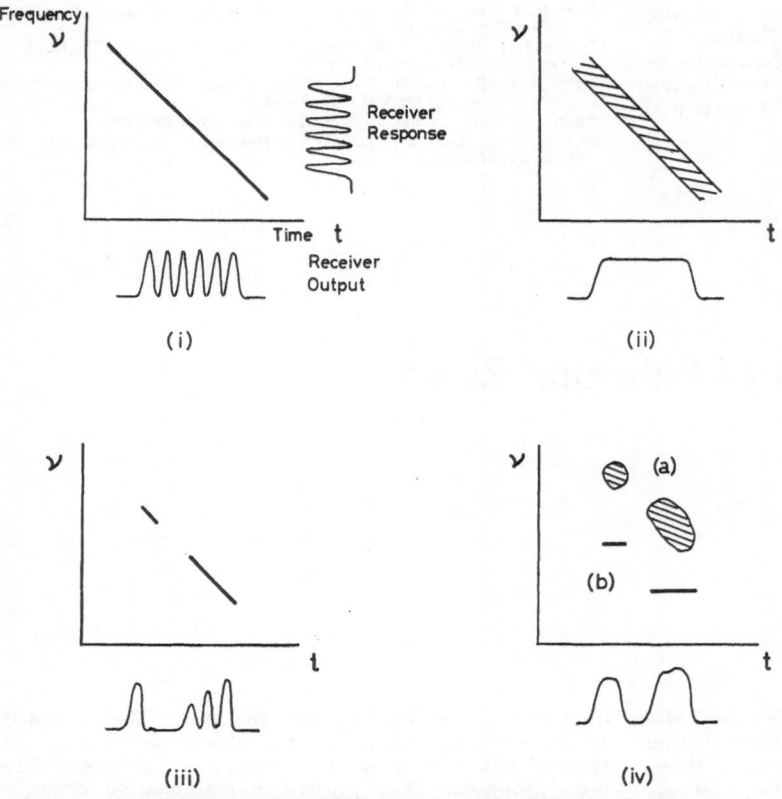

Fig. 3. The output produced by different types of received signal. (i) Narrow-band signal; (ii) wide-band signal; (iii) narrow-band signal with frequency structure; and (iv) wide-band signals with frequency structure (a) or narrow-band pulses of appreciable time duration (b).

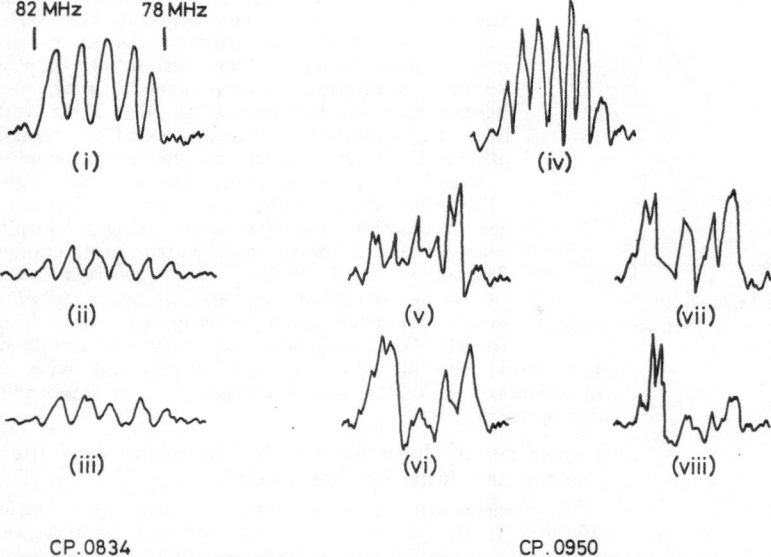

Fig. 4. Pulses observed from the sources CP 0834 (ii and iii) and CP 0950 (v–viii). Calibration pulses from a swept-frequency signal generator are also shown (i and iv).

binary model of Saslaw et al.[3] resulting from the precession of an elliptical orbit. No detectable modulation was found, the timing of the pulses during these periods being regular to within 0·02 of the pulse period in each case. These results set an upper limit of about 0·1 for the eccentricities of the orbits for both CP 0834 and CP 0950. The explanation of the sources in terms of a binary neutron star[3] thus requires either two sources of equal flux density experiencing equal amounts of gravitational focusing or, alternatively, emission from only one component of the binary; the former explanation is unlikely because of the similarity between the amplitude variations found for odd and even pulses over successive periods of 5 min. In the latter case the orbital period must be half that envisaged by Saslaw et al.[3]. In either case the orbit must be closely circular.

If, on the other hand, the radiation arises from sources of directed emission in a rapidly rotating white dwarf[4] the equal spacing of the pulses and the absence of any intermediate pulses can only plausibly be explained if the emission is produced by a single active region. The phase stability of the observed pulses during periods of 30 min and the continuity found in a day by day plot of the time of arrival of pulses[1,2] show that the same emitting region must persist for periods of several months.

The pulse to pulse amplitude variations do not seem to be explicable in terms either of interplanetary[2] or of interstellar scintillation and must be accounted for by variations within the sources themselves (unpublished results of P. A. G. Scheuer). An autocorrelation analysis of the amplitudes of successive pulses has been carried out for the sources CP 0950 and CP 1919 on the basis of observations during periods of 20 min and 80 min respectively (Fig. 2). In the case of CP 1919 it can be seen that periods of activity lasting for about 50 pulses (∼1 min) occur. The pulses from CP 0950, on the other hand, show no significant correlation for periods up to 500 pulses (2 min). The day to day variations of CP 0950 suggest, however, that a period of increased activity can persist from one day to the next.

Frequency Structure of the Pulses from CP 0950

The original observations of Hewish et al.[1] showed that the pulses from CP 1919 consisted of narrow-band signals sweeping downwards in frequency, the rate of change of frequency being about 5 MHz s⁻¹. The observations were consistent with the production at the source of short (<0·02 s) pulses of wide bandwidth which subsequently underwent dispersion in the interstellar medium. The assumption that the dispersion occurs predominantly outside the source is further supported by the measurements of Davies et al.[6], who found that on this interpretation the pulses were emitted simultaneously for a very wide range of frequencies. The measurements of three other sources[2] show that these also emit pulses with similar characteristics, although the sweep rates differ appreciably, suggesting

figures were found for CP 0834. It can also be seen that there is no evidence for any intermediate pulses with a mean amplitude exceeding 2 per cent of that of the main pulses.

The timing of the pulses from CP 0834 and CP 0950 has been investigated over periods of 20–30 min in order to detect any phase modulation of the pulses occurring with periods of between 5 min and 60 min. A modulation with a period in this range would be expected on the that the ray path includes different values of integrated electron density.

Further observations have now been made to investigate the frequency structure of the source *CP* 0950. In these observations the signal from one half of the aerial was delayed by means of an additional cable length, l, and combined with the signal from the other half of the aerial before being amplified and recorded as described earlier. The response of the system as a function of signal frequency, ν, was thus a series of interference fringes of the form

$$\cos^2\left(\frac{\pi c}{l\nu}\right) \text{ where } c \text{ is the velocity of light}$$

These fringes extend over the 4 MHz bandwidth of the receiver, which was centred on 80·0 MHz. The additional cable length, l, was 440 m, so that successive maxima were separated by 0·6 MHz. The system was calibrated by injecting into the cables from each half of the aerial small signals from a signal generator which could be swept in frequency.

The form of the output for several possible types of input signals is illustrated in Fig. 3.

Signals having an instantaneous bandwidth greater than about 300 kHz will smooth out the interference fringes in the frequency response of the system, while any frequency structure present in the original pulse will, after dispersion, appear as a modulation of the receiver output in time. The type of output shown in Fig. 3 (iv) can be explained either in terms of frequency components showing a wide instantaneous bandwidth (> 300 kHz) or, alternatively, in terms of narrow-band frequency components of longer duration; on either interpretation the original pulse must be emitted over a time interval greater than about 15 ms.

Fig. 4 shows two pulses from the source *CP* 0834 (ii and iii) together with a calibration pulse from a swept-frequency signal generator (i). The pulses are completely consistent with a short emitted pulse of wide bandwidth, as found earlier in the case of *CP* 1919, and show no evidence for complex frequency structure. The pulses observed from *CP* 0950, on the other hand, show a wide variety as shown in Fig. 4 (iv to viii). It is immediately apparent that the pulses cannot generally be interpreted in terms of a simple swept-frequency signal

such as would arise from a short emitted pulse of wide bandwidth. Frequency structure on a scale of $\leqslant 1$ MHz usually appears to be present, the same structure often being repeated over periods of up to 1 min. The absence of interference fringes in most of the pulses implies a time of emission which is usually greater than about 15 ms.

It is possible that some of the effects observed might be explained in terms of Faraday rotation in the interstellar medium if the emission from *CP* 0950 contains a linearly polarized component; alternatively they may be caused by interstellar scintillation. The observed dispersion in the case of *CP* 0950 implies an integrated electron density, along the line of sight, of $3\cdot2 \pm 0\cdot8$ cm^{-3} pc (ref. 2). A value of the mean component of magnetic field along the line of sight of 3×10^{-6} gauss would be sufficient to produce a rotation of the plane of polarization of ~ 5 radians across the 4 MHz band of the receiver. It is therefore possible that some of the observed frequency structure could be produced in this way although it does not explain the variability from pulse to pulse.

Scheuer (unpublished results) has examined the possibility of accounting for the longer period variations of pulse amplitude in terms of scintillation caused by irregularities of electron density in the interstellar medium. A scintillation mechanism might also account for the structure in the spectrum of the individual pulses, but seems unlikely to explain the rapid (< 5 min) variations observed.

It therefore seems necessary to suppose that at least part of the fine frequency structure observed in the pulses of *CP* 0950 arises within the source itself.

We are indebted to Sir Martin Ryle and Dr P. A. G. Scheuer for helpful discussions and one of us (R. A. C.) thanks the Science Research Council for a maintenance award.

Received April 5, 1968.

[1] Hewish, A., Bell, S. J., Pilkington, J. D. H., Scott, P. F., and Collins, R. A., *Nature*, **217**, 709 (1968), (Paper 1).
[2] Pilkington, J. D. H., Hewish, A., Bell, S. J., and Cole, T. W., *Nature*, **218**, 126 (1968), (Paper 2).
[3] Saslaw, W. C., Faulkner, J., and Strittmatter, P. A., *Nature*, **217**, 1222 (1968), (Paper 42).
[4] Ostriker, J., *Nature*, **217**, 1227 (1968), (Paper 39).
[5] Ryle, M., *J. Inst. Elect. Eng.*, **6**, 14 (1960).
[6] Davies, J. G., Horton, P. W., Lyne, A. G., Rickett, B. J., and Smith, F. G., *Nature*, **217**, 910 (1968), (Paper 6).

9. Detection of the Pulsed Radio Source *CP* 1919 at 13 cm Wavelength

by
A. T. MOFFET
R. D. EKERS
Owens Valley Radio Observatory,
California Institute of Technology,
Pasadena, California

THE pulsed radio source discovered at 3·68 m wavelength[1] has been studied at various other wavelengths between 2 m and 73 cm[2],[3]. The latter report states that the individual pulses have widths of about 35 ms with a sharp leading edge and with some indication of a second peak near the end of the pulse. Peak flux densities of about 50 flux units are reported by both groups, with time-average fluxes of a few tenths of a flux unit (that is, a few times 10^{-27} Wm^{-2}Hz^{-1}). At shorter wavelengths, upper limits of

0·1 flux unit were obtained by Ryle and Bailey at 21·4 cm and by Davies *et al.* at 11·1 cm.

On March 16 and 17 we had an opportunity to observe this source at 13 cm wavelength with extremely sensitive equipment at the Goldstone Tracking Station of the Jet Propulsion Laboratory. Observations were made with a maser receiver on the 210 ft. (64 m) Advanced Antenna system. The total noise temperature present in this receiving system when pointed at "cold sky" near the zenith was 20° K. Because the pulsed source is in the galactic plane and the observations were made at rather large zenith angles, the system temperature during the observations was about 26° K. The maser amplifier was followed by a dual-conversion superheterodyne receiver with intermediate frequencies of 50 and 10 MHz; the receiver accepted a band 6·1 MHz wide centered at 2,295 MHz. The detector was followed by several stages of direct current amplification having a smoothing time constant of 0·002 s. Data were sampled 1,000 times a second by an analogue-to-digital converter and were recorded in digital form on magnetic tape under the control of a small computer. A flux scale was established through observations of the sources 3C17 and 3C48. The feed accepted linear polarization with the electric vector in approximately position angle 70°.

The data were subsequently analysed with a digital computer in which the functions of a conventional signal averager were simulated. The expected pulse period was divided into 1,337 cells of approximately 1 ms duration. The time of each data sample was calculated *modulo* the expected period, and the sample was added to the appropriate cell. Each minute of data was treated separately in this manner, and the successive 1 min averages were examined for the presence of a recurring pulse with the correct period; such a pulse would be found in the same cell in each 1 min average.

The expected period used in the analysis was 1.337200 s. This value is close to that which would be observed for a source with a period of 1·3372795 s in the reference frame of the Sun, as reported by Hewish *et al.* For our observations, the apparent period (adjusted for the Earth's orbital motion, including ellipticity, and for its rotation) would be 1·3371972 s on March 16 and 1·3371965 s on March 17.

The results of our observations on March 16, which covered the periods 2037 to 2040 and 2041 to 2101 UT, are shown in Fig. 1. Each trace represents a 5 ms average

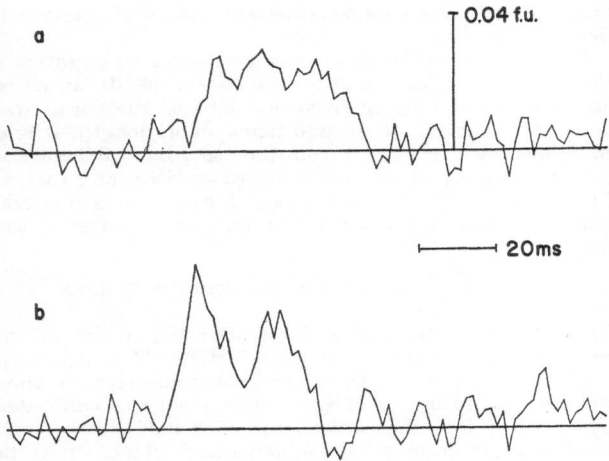

Fig. 2. *a*, Twenty-three minute average of the received signal assuming the true period is that given in ref. 1. *b*, A similar 23 min average assuming the apparent period is 1·3372223 s.

within the expected period and a further average over 4 min of observation (except for the last, which is 3 min). As can be seen, the pulse appears in each of these traces. It has a width of about 40 ms with peaks near each end of this interval. As time progresses (from top to bottom in Fig. 1), the pulse drifts towards the right. While it is possible that this apparent drift is spurious (caused, perhaps, by the poor statistics of detection in the 4 min averages), it seems possible that it is real. This would indicate a period during the time of these observations which was $2·6 \times 10^{-5}$ s longer than that reported by Hewish *et al.* It is not clear whether other reported measurements of the pulse period would have been sensitive to short term variations in pulse arrival time of a few hundredths of a second. Davies *et al.* remark that the arrival time of the pulses at 408 MHz is stable to within 5 ms, but it is not clear whether they compared this with the expected arrival time calculated on the basis of the period reported by Hewish *et al.* or on the basis of the mean period of a long sequence of pulses.

The peak flux density of any of the 5 ms resolution 4 min averages shown in Fig. 1 is 0·065 flux units. The peak in any of the 2 ms resolution 1 min averages is 0·11 flux units. The standard deviations in these averages due to noise fluctuations are 0·011 and 0·031 flux units, respectively.

The average shape of the pulse was obtained by summing all twenty-three 1 min averages. This was done in two ways: first, assuming that the correct period was that reported by Hewish *et al.* (that is, by summing along vertical lines in Fig. 1); and second, assuming an apparent period of 1·337225 s, which approximately aligned the leading peaks (that is, by summing along lines parallel to the inclined line shown in Fig. 1). The results are shown in Fig. 2. As might be expected, the second procedure gave an average pulse with more sharply defined structure. If the second procedure is correct, the average pulse had a width of about 40 ms, with a sharp leading peak about 12 ms wide. A representative time for reception of this initial peak is 203700·86 UT, March 16. The peak flux density is 0·040 flux units. The average flux density for the 40 ms duration of the pulse is $0·024 \pm 0·002$ flux units, and the long-time average is then 40/1337 of this value, or 0·00072 flux units. These flux densities are several hundred times smaller than those reported at lower frequencies. Incidentally, the average flux density is at least one order of magnitude fainter than for any previously detected radio source.

The drop in intensity at high frequencies is extremely steep. If the average flux density observed on March 16 is in any sense typical, and if the same is true of the

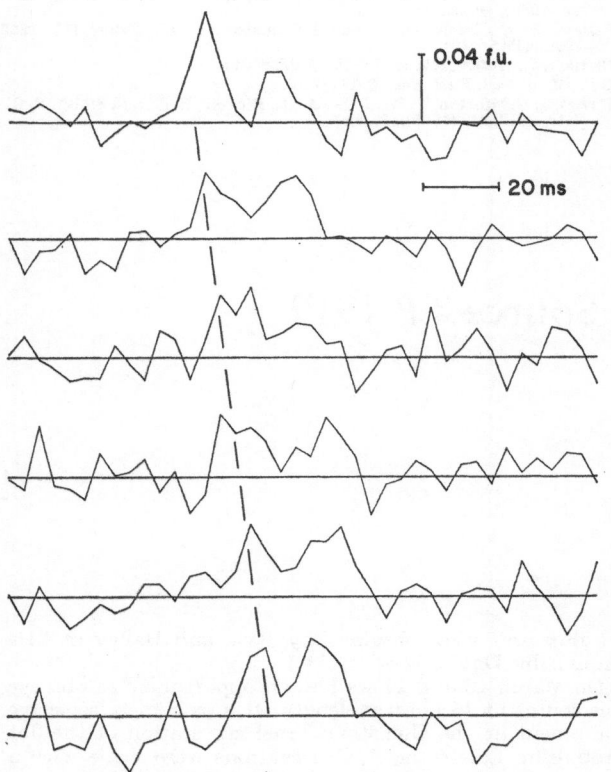

Fig. 1. Four minute averages of the received signal from *CP* 1919. Time within the pulse period increases towards the right. The inclined dashed line represents an apparent pulse period of 1·3372223 s.

average 73 cm flux reported by Ryle and Bailey, the radio spectral index between these two wavelengths is −3·2.

A significant feature of this observation is that the pulses have the same width at 13 cm as at 73 cm. Indeed, the similarity between our Fig. 2b and Fig. 1b of Davies et al. is quite striking. At the shorter wavelength, propagation effects should be negligible, indicating that this width is an intrinsic property of the source. The observed width should not be affected by dispersion within our receiver bandwidth. If the dispersion follows the wavelength-squared law reported by Davies et al., the delay time is only about 10^{-2} s from zero wavelength to 13 cm; the time to cross our receiver band should be of the order of 10^{-4} s.

Similar observations made between 1929 and 2000 UT on March 17 gave no convincing evidence for pulses during that time. The level of detection on March 17 was slightly poorer than on March 16 because of data sampling errors; nevertheless, pulses with the shape shown in Fig. 1 should have been detected if they were one-quarter the strength observed on March 16. Thus the reported variation in intensity from day to day would seem to persist at short wavelengths. Although our signal to noise ratio is not high enough to make any detailed comment on short period time variations, no pulse was seen in three 1 min averages, and this variation is probably real.

We thank J. R. Hall, B. Seidel and S. Anastos of the Jet Propulsion Laboratory for their assistance with these observations. We also thank the Goldstone operating personnel for their co-operation. One of us (A. T. M.) was assisted by an Alfred P. Sloan research fellowship. The programme of research in radio astronomy at the California Institute of Technology is supported by the US Office of Naval Research.

Received April 16, 1968.

[1] Hewish, A., Bell, S. J., Pilkington, J. D. H., Scott, P. F., and Collins, R. A., Nature, 217, 709 (1968), (Paper 1).
[2] Ryle, M., and Bailey, J. A., Nature, 217, 907 (1968), (Paper 19).
[3] Davies, J. G., Horton, P. W., Lyne, A. G., Rickett, B. J., and Smith, F. G., Nature, 217, 910 (1968), (Paper 6).

10. Measurements of the Pulse Shape and Spectra of the Pulsating Radio Sources

by
A. G. LYNE
B. J. RICKETT
University of Manchester,
Nuffield Radio Astronomy Laboratories,
Jodrell Bank

FOLLOWING the observations[1,2] of the pulsating radio source CP 1919 at $\alpha = 19^h 19^m$, $\delta = 21° 47'$, the three other sources reported by Hewish et al.[1] have been studied with the 250 ft. telescope at Jodrell Bank. Simultaneous observations have been made on frequencies of 151 MHz, 408 MHz and 922 or 1,412 MHz, to obtain the spectrum of the radio emission and to compare the shapes of the pulses emitted by the four known sources.

Preliminary values for the positions of these sources were kindly provided by Dr A. Hewish. Improved values for the declinations were obtained by repeated scans across the sources at 408 MHz, although there may still have been sufficient error in telescope pointing to affect the observations at the highest frequencies. The best estimates of the positions are shown in Table 1.

Table 1

	α (1950)	δ (1950)
CP 0834	08h 34m 07s ± 15s	06° 18′ ± 10′
CP 0950	09h 50m 28s	08° 09′ ± 8′
CP 1133	11h 33m 32s ± 20s	16° 09′ ± 10′

The right ascensions are the values provided by Cambridge. All our measurements used these positions.

We report measurements on the dispersions, pulse shapes, amplitude variations and spectra of these three sources together with the earlier measurements on CP 1919 already published[2]. The same techniques and analysis procedures have been used.

Pulse Delay

The delay in arrival time of the pulses has been attributed to dispersion in propagation time through ionized interstellar gas, giving a delay proportional to (frequency)$^{-2}$. We tested this law by plotting the difference in arrival times $t_1 - t_2$ at a pair of frequencies f_1 and f_2, against $(1/f_1^2) - (1/f_2^2)$ as in Fig. 1. Ambiguities in the whole number of periods in the delay were eliminated by observing at 150·5 MHz and 151·5 MHz simultaneously.

All the sources give straight line plots, with an intercept on the vertical axis of 0 ± 5 ms. This suggests that the emission takes place within 5 ms at frequencies between 150 MHz and 922 MHz for all four sources.

From the slopes of the straight lines we obtain the following values for the integrated electron content in the line of sight for the four sources. An estimate of the distance of the sources assuming a mean electron density of 0·1 cm^{-3} is also given in Table 2.

Table 2.

	$N\,dl$ (pc cm^{-3})	Distance (pc)
CP 0834	12·80	128
CP 0950	2·94	30
CP 1133	4·87	49
CP 1919	12·55	126

Pulse Shapes

High resolution measurements were made both on individual pulse shapes using fast galvanometer recordings and on the pulse shapes integrated in an on-line computer over a period of a few minutes. The pulses are broadened by dispersion across the receiver bandwidth. This broadening has been limited in all these observations to a few ms by restricting the receiver bandwidth.

The pulses have been shown to contain components which are highly linearly polarized[3] and hence the detailed pulse shapes observed are likely to be strongly dependent on the aerial polarizations. During most of the measurements, the 151 MHz polarization was circular, giving a correct measure of pulse shape, but the 408 and 922 MHz

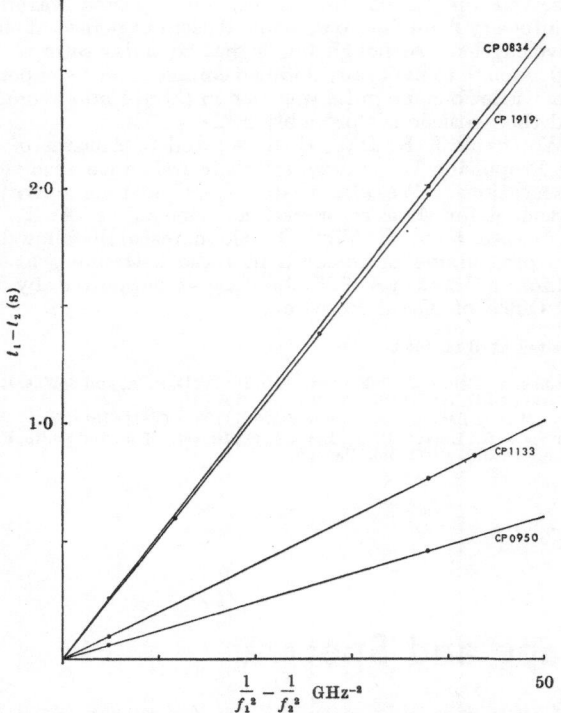

Fig. 1.   The dispersion in arrival time of the pulses for the four sources.
Measurements were made at 151, 240, 408 and 922 MHz.

source. In fact, each profile can be represented by the sum of two gaussian components of width 17 ± 4 ms and separation 20 ± 4 ms. It may be significant that the ratio of the amplitudes of these two components varies monotonically with repetition period of the pulses.

The double nature of the pulses suggests two distinct emitting regions separated by 6,000 km in the direction of the line of sight. This interpretation is substantiated by the observation that the two components for the source CP 1133 have different polarization angles[3]. This requires the magnetic field to be different at the origins of the two components, which by implication must be separate but which must nevertheless be emitting with the same spectrum.

Pulse Amplitude Variations

All four sources show large variations in the amplitude of the pulses, both from pulse to pulse and on a longer time scale. To allow comparison between observations using different receiver time constants and bandwidths, we have used values of the pulse energy density, the energy received per pulse in J Hz⁻¹ m⁻². Galvanometer recordings were used to examine variations from pulse to pulse, while longer period variations were sought in values of pulse energy averaged over 2 min.

Because of the polarization of the pulses, the 408 MHz and 922 MHz amplitudes are subject to variations due to changes in polarization angle caused by the ionosphere

aerials were linearly polarized and some individual pulse shapes may have been affected by this.

Some typical sequences of pulses for the sources CP 0950, CP 1133 and CP 1919 are illustrated in Fig. 2. These galvanometer recordings were made during fairly active periods. No records exist for CP 0834, which is usually weaker than the other three. At 408 MHz all three sources show pulses which have the same general characteristics. Individual pulses show power emission lasting for between 5 and 50 ms. During this period there are often several distinct features some of which are unresolved by the recording system and have durations of less than 1 ms. These individual features sometimes survive for several pulses or they may appear in only a single pulse.

Simultaneous observations of pulse shapes on up to three frequencies have been made on CP 0950, CP 1133 and CP 1919. Fig. 4a shows one record of two successive pulses observed at 151, 408 and 922 MHz on CP 0950. The pulse shape appears to be the same on the three frequencies. The data on CP 1133 and CP 1919, while limited by sensitivity, agree with this observation.

Although the individual pulse profiles are very variable from one pulse to the next, each source does appear to have a well defined mean pulse shape when averaged over a few minutes. These mean profiles show no significant change during our observations. Typical mean pulse shapes for each of the four sources at 408 MHz are displayed in Fig. 3. The effect of changing polarization across the pulse has been removed as far as possible by making observations at several different polarization angles. The total durations of the mean pulse profiles shown in Fig. 3 are all in the range 50 to 60 ms, after removal of the effects of dispersion in the receiver bandwidth and the finite resolution of the integration procedure. Three of them appear to have a distinctly double nature, while the fourth, CP 0950, has a marked extended feature before the main peak, which could be a second much weaker component. This is supported by the occasional appearance, on the individual pulse recordings, of a preliminary pulse some 30 ms before the main pulse. Fig. 2 shows an example of this in some of the pulses from this

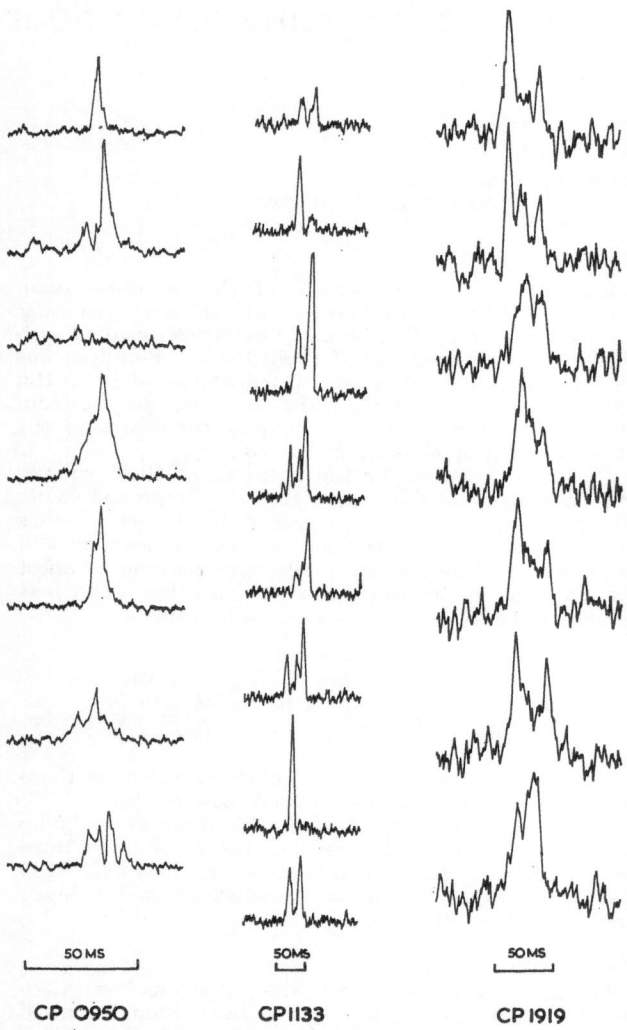

CP 0950   CP1133   CP 1919

Fig. 2.   Series of consecutive pulses recorded directly at 408 MHz.
(Bandwidth 4 MHz time constant 1 ms.)

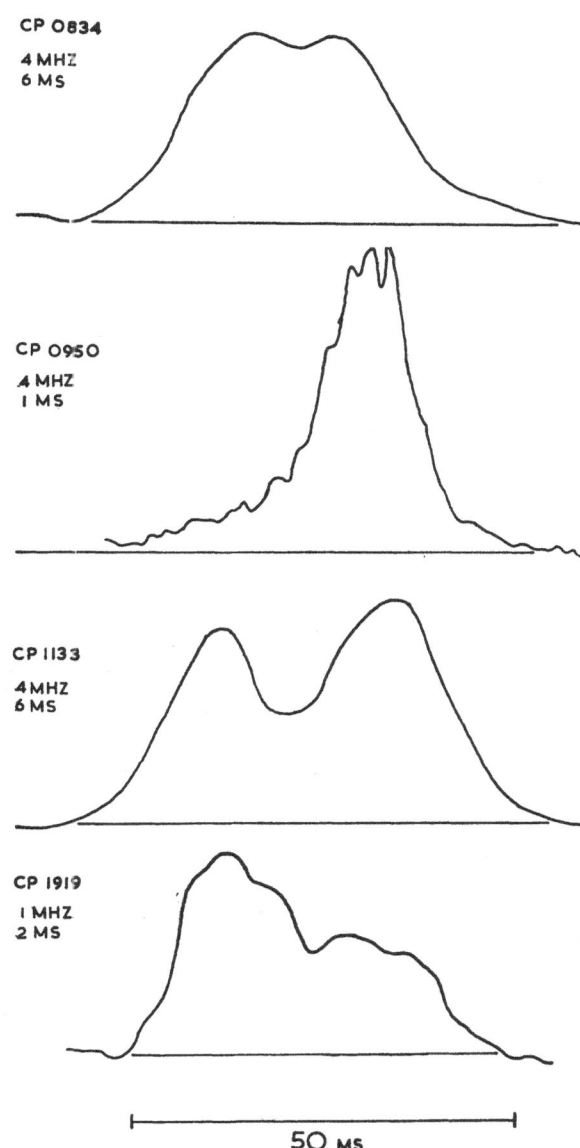

CP 0834
4 MHZ
6 MS

CP 0950
4 MHZ
1 MS

CP 1133
4 MHZ
6 MS

CP 1919
1 MHZ
2 MS

50 MS

Fig. 3. Mean pulse profiles averaged over 8 min at 408 MHz for the four sources. The receiver bandwidths and effective time resolutions are indicated.

and the Earth's rotation. Neither of these effects is sufficient to account for the observed amplitude variations. A possibility which cannot be ruled out is that there is a change in the polarization angle within the sources over periods of several hours.

Groups of pulses are often observed for three of the sources at both 151 MHz and 408 MHz, showing that the pulse amplitude is correlated over only a few pulses. An apparent exception is CP 1919 at 408 MHz, where correlation exists over about 20 pulses. We have no useful measurements of the fourth source CP 0834 because it seems to be somewhat weaker and individual pulses were seldom seen.

CP 0950 shows clear correlation of the pulse to pulse amplitude pattern at the three frequencies used. Thus a large pulse at 408 MHz can be identified at 151 MHz and 922 MHz, providing, incidentally, unambiguous confirmation of the relative dispersion delays. The relative amplitudes at these frequencies remain constant for a few minutes. This implies that the spectrum of the source remains constant over a few minutes, and that the short term variations are intrinsic to the source.

The longer term variations are shown in Fig. 5 for each of the four sources over a period of a few hours. At first glance there is little correlation between variations on the different frequencies. Over a few minutes the variations do show some correlation, however, confirming that the spectrum is constant over this time scale. On the other hand, there is little correlation over periods of 30 min. This could be explained in terms of a changing spectrum within the source or possibly by deep scintillations in the interstellar medium.

Spectra

For the purpose of elucidating emission mechanisms, a mean spectrum can be defined for each source. Fig. 6 shows the pulse energy density averaged over the periods of observation of Fig. 5 for each source, as a function of frequency. These spectra are subject to possible errors in the source positions, the result of which would be a reduction in the observed power at the higher frequencies. We have no clear evidence for a cut-off in these spectra, though there is a steady decline with increasing frequency.

Conclusion

In spite of a wide variety of pulse shapes and amplitudes at different times and from different sources, some general characteristics have emerged. After removing the relative dispersion delays we find that the time of emission is the same, within 5 ms, over a wide frequency range (151–922 MHz). Although the repetition periods

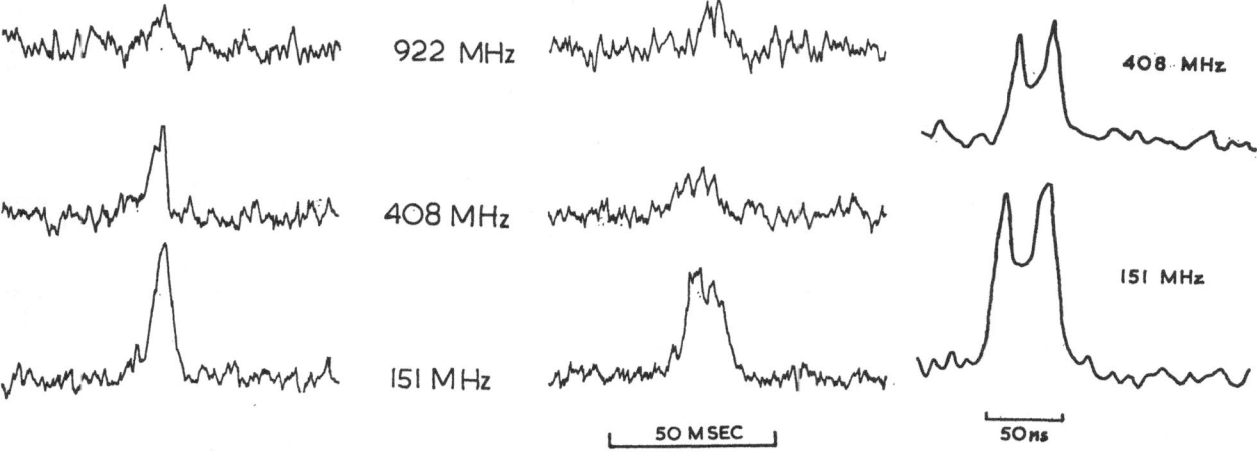

922 MHz

408 MHz

151 MHz

408 MHz

151 MHz

50 MSEC

50 MS

Fig. 4. a, Two consecutive pulses on CP 0950 observed on 922 MHz, 408 MHz and 151 MHz. The relative dispersion delays have been removed. b, Mean pulse profiles averaged over 2 min on CP 1133 at 151 MHz and 408 MHz.

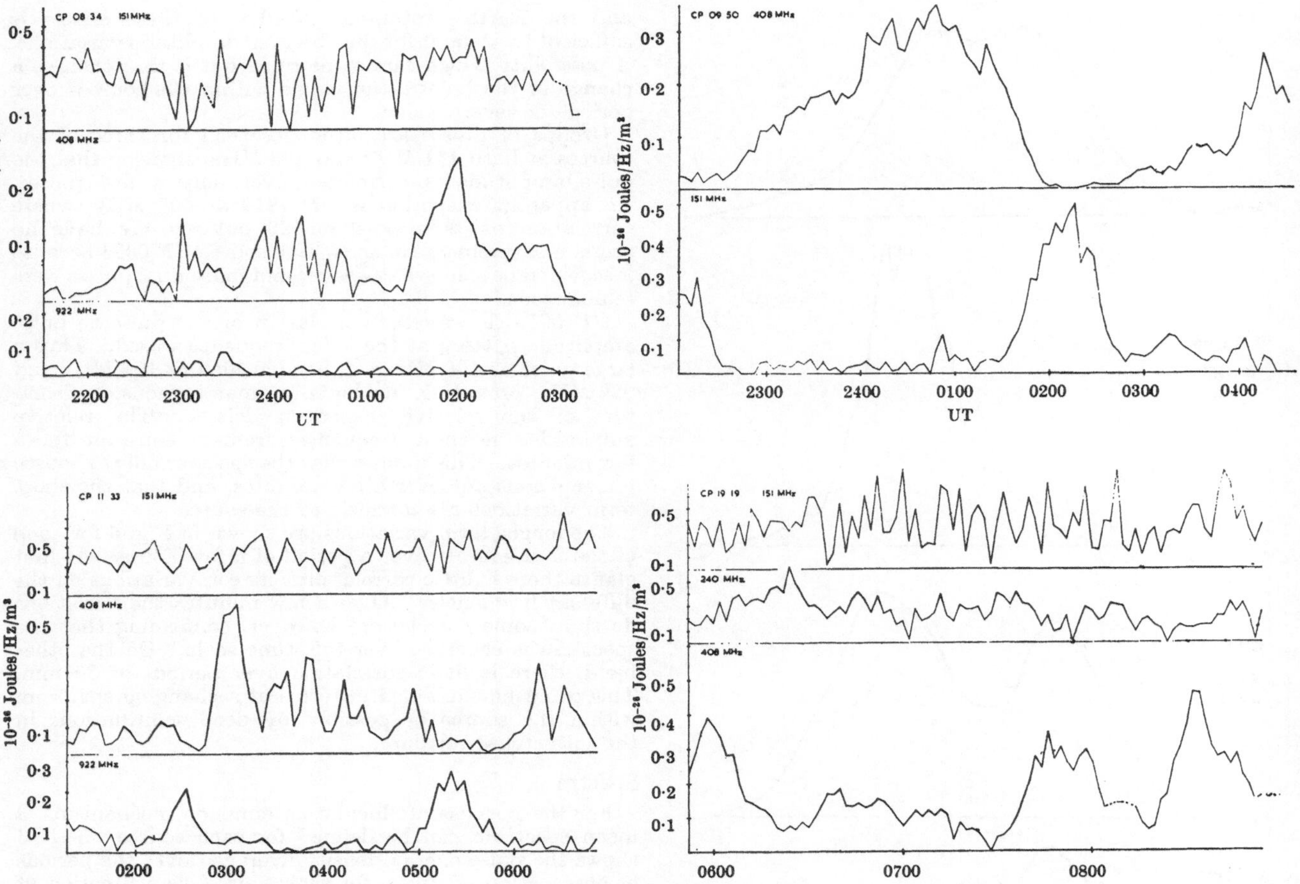

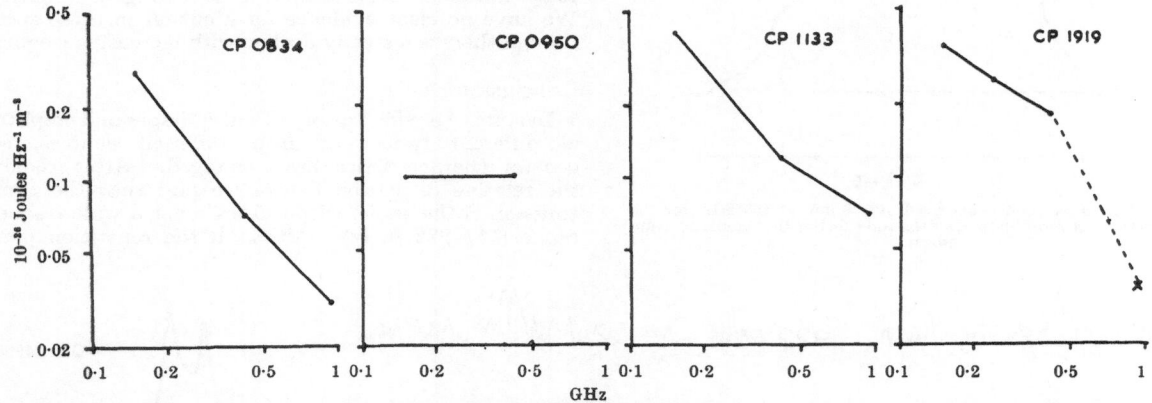

Fig. 5. Simultaneous amplitude observations at two or three frequencies for the four sources. The amplitudes are the mean pulse energy densities averaged over 2 or 4 min.

Fig. 6. Mean pulse energy density as a function of frequency taken over the periods of observation of Fig. 5. X is a measurement made on a different occasion.

are extremely constant, the amplitude and detailed shape of the pulses vary over only a few seconds. The pulse shapes are the same on all frequencies and show instantaneous fine structure down to a few ms, but have mean shapes with total durations about 50 ms. The mean shapes are characteristic of each source and have a tendency to be double.

We have measured the emission in the interval between pulses, and found it to be less than 1 per cent of peak pulse emission during an outburst on *CP* 0950. It seems hard for the gravitational focusing theory[4] to explain the amplification factor of more than 100 implied by this measurement.

Because of the very high brightness temperatures of 10^{21} °K implied by our measurements, the emission process must be coherent. To account for emission over a very broad band, the radiation at any instant must originate in a slab only 10 cm thick. This may be a shock front or layer of particles passing through an emitting region or regions the properties of which determine the detailed pulse shape and amplitude.

Received April 18, 1968.

[1] Hewish, A., Bell, S. J., Pilkington, J. D. H., Scott, P. F., and Collins, R. A., *Nature*, **217**, 709 (1968), (Paper 1).

[2] Davies, J. G., Horton, P. W., Lyne, A. G., Rickett, B. J., and Smith, F. G., *Nature*, **217**, 910 (1968), (Paper 10).

[3] Lyne, A. G., and Smith, F. G., *Nature*, **218**, 124 (1968), (Paper 7).

[4] Saslaw, W. C., Faulkner, J., and Strittmatter, P. A., *Nature*, **217**, 1222 (1968), (Paper 42).

11. Further Impulsive Emission from *CP* 0950

by

B. J. RICKETT
A. G. LYNE
University of Manchester,
Nuffield Radio Astronomy Laboratories,
Jodrell Bank.

DURING observations of the polarization of the impulsive emission from the source *CP* 0950 a further feature of the profile reported previously[1,2] has become apparent. Using the integration technique described earlier[3], observations suggested the presence of impulsive emission in the region between the main pulses. Fig. 1 is a mean profile obtained at 408 MHz using the Mark I telescope. The newly discovered "interpulse" is evident and lies 100 ms before the peak of the main pulse. Subsequent measurements have demonstrated that this interpulse is very highly polarized.

Simultaneous measurements were made with two receivers connected to orthogonal dipoles at the focus of the Mark I telescope. As the feed was rotated, the interpulse disappeared from one of the two receivers and reappeared in the other. Fig. 2 shows the results of 50 min simultaneous integration on the interpulse on the two orthogonally polarized receivers. The position angle of the polarization of the interpulse nearly coincides with that of the receiver producing the upper trace. During these observations we have estimated that the interpulse has between 85 and 100 per cent of its radiation at 408 MHz linearly polarized. At the same time, the peak of the main pulse was only about 15 per cent linearly polarized. The duration of the interpulse is about the same as that of the main pulse. The energy in this feature represents about $1 \cdot 8 \pm 0 \cdot 4$ per cent of the energy in the main pulse. When the main pulse is weaker, the interpulse disappears below the noise threshold of the observations. Our results are consistent with a constant ratio of the energy densities in the interpulse and main pulse. Observations have also been made at 151 MHz and these show the existence of a similar feature, again placed 100 ms before the peak of the main pulse. The ratio of the energy densities in the interpulse and main pulse seems to be similar to that observed at 408 MHz.

The sources *CP* 1133 and *CP* 1919 have also been investigated for emission outside the main pulse, but no impulsive feature of comparable magnitude with the interpulse in *CP* 0950 has been detected at 408 MHz. In *CP* 1133 there is a single possible feature 400 ms before the main pulse having $0 \cdot 6 \pm 0 \cdot 4$ per cent of the energy density in the main pulse. In *CP* 1919 there is a possible feature 460 ms before the main pulse of $0 \cdot 5 \pm 0 \cdot 3$ per cent. These are only marginally significant and must be confirmed by further measurements.

The origin of the interpulse must be similar to that of the main pulse, as the duration and the spectrum are both the same. The differences in polarization suggest, however, that the two pulses do not represent the repeated excitation of a single emitting region, as when a second shock front crosses a single sharp boundary in a magneto-ionic medium. We suggest that there are on the source two emitting regions separately excited by related shock fronts. The delay in excitation, by 100 ms or by 153 ms according to the sequence of the two pulses, might be determined by the velocity of a single shock front arriving at different times in the two regions. Because the delay is the same within 10 ms at the two observing frequencies,

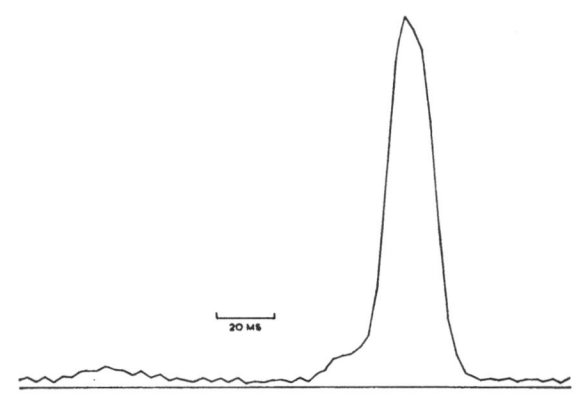

Fig. 1. *CP* 0950, May 2, 1968, 0 h UT. Mean pulse profile at 408 MHz.

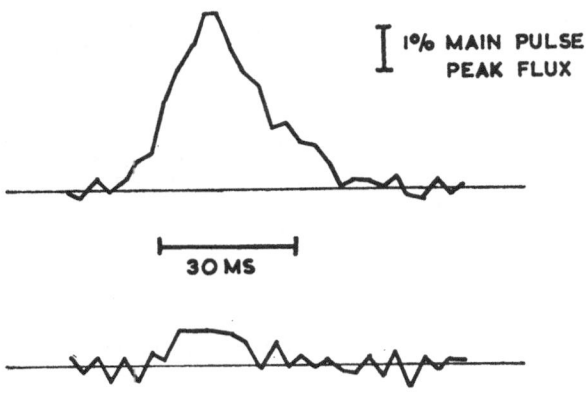

I 1% MAIN PULSE PEAK FLUX

30 MS

Fig. 2. *CP* 0950. Two profiles of the interpulse observed on orthogonal linear polarizations at 408 MHz.

the lines of sight to these two regions must contain the same integrated electron content within 2×10^{12} cm^{-3} km.

The presence of impulsive radiation outside the main pulse seems to be inexplicable on the gravitational focusing model of Saslaw, Faulkner and Strittmatter[4]. On the model of a rotating white dwarf or neutron star[5,6], the interpulse must be interpreted as a second emitting region on the object the average energy output of which must be a constant fraction of the output of the main emitting region.

Received May 29, 1968.

[1] Lyne, A. G., and Rickett, B. J., *Nature*, **218**, 326 (1968), (Paper 10).
[2] Lyne, A. G., and Smith, F. G., *Nature*, **218**, 124 (1968), (Paper 7).
[3] Davies, J. G., Horton, P. W., Lyne, A. G., Rickett, B. J., and Smith, F. G., *Nature*, **217**, 910 (1968), (Paper 6).
[4] Saslaw, W. C., Faulkner, J., and Strittmatter, P. A., *Nature*, **217**, 1222 (1968), (Paper 42).
[5] Ostriker, J. P., *Nature*, **217**, 1227 (1968), (Paper 39).
[6] Gold, T., *Nature*, **218**, 731 (1968), (Paper 40).

12. Observations of Pulsating Radio Sources at 11 cm

by

E. J. DAINTREE
J. G. DAVIES
P. W. HORTON
D. WALSH

University of Manchester,
Nuffield Radio Astronomy Laboratories,
Jodrell Bank.

THE four known pulsating radio sources[1,2] have all been observed at wavelengths down to 32 cm (ref. 3). For the source CP 1919 an upper limit to the flux density at 21 cm (ref. 4), a positive detection at 13 cm (ref. 5), and an upper limit at 11 cm (ref. 6) have been published. The previous observations at 11 cm wavelength have now been extended to all four sources, and positive results are reported here for two of them.

The observations were made with the Mark 2 telescope at 2,695 MHz on May 8, 9 and 10, 1968. The receiver bandwidth was 8 MHz, the system temperature 170° K and the aperture efficiency 50 per cent. The observing technique, using superposition of the receiver output over a number of cycles of the pulse repetition period by means of a computer, has been described previously[6]. Observations were averaged for 80 min on CP 0950 and for 60 min on each of the other sources.

Pulses from CP 0950 and CP 1133 were clearly detected. Their arrival times agreed with those expected from comparison with simultaneous observations at 408 MHz using established dispersion delays[3,6]. Fig. 1 illustrates this for CP 0950. The signal-to-noise ratio was not sufficient to determine definite pulse profiles, but the durations appeared similar to those reported at lower frequencies[3]. For CP 0834 and CP 1919 only upper limits could be placed on 11 cm emission. Details of the average pulses are given in Table 1. Mean pulse energy is estimated by assuming an effective pulse duration of 30 ms at peak flux density.

Table 1. DETAILS OF THE FOUR SOURCES

Source	Peak flux density (W m⁻² Hz⁻¹)	Mean pulse energy (J m⁻² Hz⁻¹)
CP 0834	$\leqslant 0\cdot1 \times 10^{-26}$	$\leqslant 3 \times 10^{-29}$
CP 0950	$0\cdot15 \times 10^{-26}$	$4\cdot5 \times 10^{-29}$
CP 1133	$0\cdot3 \times 10^{-26}$	9×10^{-29}
CP 1919	$\leqslant 0\cdot1 \times 10^{-26}$	$\leqslant 3 \times 10^{-29}$

CP 1919 would not be expected to be above the detection limit of either these or the previous observations[6] at 11 cm wavelength, assuming the mean pulse energy at 13 cm is 10^{-29} J m⁻² Hz⁻¹ (ref. 5). This source apparently has a spectral index between 73 cm and 13 cm of $3\cdot2$ (ref. 5). If this spectral index were characteristic of all the sources over this wavelength range, then none should have been detected at 11 cm. In fact, for both CP 0950 and CP 1133 the spectral index over this range is not significantly different from $1\cdot5$. For CP 0834 the index is steeper than $1\cdot7$. Because of the variability of the sources, there is some inherent uncertainty in these figures which assume the 11 cm results and the previous longer wave-

length results[3] are representative. It seems likely, however, that the steep spectral index of CP 1919 is by no means typical of all pulsed radio sources, and there may indeed be a considerable range of spectral indices including ones flatter than $1\cdot5$. Searches at long wavelengths are biased in favour of sources with steep spectral index, and it is possible that some pulsed sources may be more amenable to detection at short wavelengths, as is the case with some known steady radio sources.

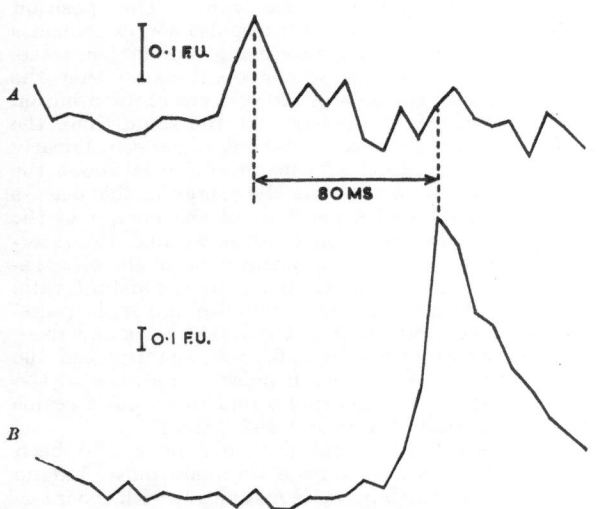

Fig. 1. CP 0950. A, 2,695 MHz data integrated 80 min. B, 408 MHz data integrated 5 min. Expected dispersion delay, 75 ms. Effective time resolution, 8 ms.

It is interesting to note that with the sensitivity reported at 13 cm wavelength[5], individual pulses from CP 1133 may just be detectable. This would be particularly interesting as a test of possible mechanisms for producing the fluctuations characteristic of pulsed radio sources at longer wavelengths.

Received May 23, 1968.

1 Hewish, A., Bell, S. J., Pilkington, J. D. H., Scott, P. F., and Collins, R. A., *Nature*, 217, 709 (1968), (Paper 1).
2 Pilkington, J. D. H., Hewish, A., Bell, S. J., and Cole, T. W., *Nature*, 218, 126 (1968), (Paper 2).
3 Lyne, A. G., and Rickett, B. J., *Nature*, 218, 326 (1968), (Paper 10).
4 Ryle, M., and Bailey, J. A., *Nature*, 217, 907 (1968), (Paper 19).
5 Moffet, A. T., and Ekers, R. D., *Nature*, 218, 227 (1968), (Paper 9).
6 Davies, J. G., Horton, P. W., Lyne, A. G., Rickett, B. J., and Smith, F. G., *Nature*, 217, 910 (1968), (Paper 6).

13. Preliminary Results on Pulsating Radio Sources

by

G. GRUEFF
G. ROFFI
M. VIGOTTI

Laboratorio Nazionale di Radioastronomia,
Istituto di Fisica,
Università di Bologna.

A PROGRAMME of observations of the pulsating sources recently discovered by the Cambridge group has been started at 408 MHz with the N.–S. arm of the "Northern Cross" radio telescope. At present only this arm of the Cross can be used because the E.–W. arm is being re-conditioned. Some of the results obtained, however, show that even these incomplete observations may be useful, because of the long transit time (4 min) in the antenna fan beam which allows the observation of hundreds of pulses at each transit and because of the multi-beam arrangement which enables a rapid and precise determination of the declination of a source to be made. Indeed, by observing the source with the five beams of the instrument simultaneously—at these declinations each beam is 9 min of arc wide and spaced 4 min of arc from the next—it is possible, in principle, to obtain a value of δ even for a single recorded pulse. Because of the very great intensity variations of pulsating sources, this technique is advantageous as compared with the conventional technique of scanning across the source.

Up to now three sources—namely CP 0950, CP 1133 and CP 1919—have been observed, each during several transits. CP 0834 has not yet been observed; we do not know whether this is only because the published position is too widely different from the true one.

The following values have been obtained:

$$CP\ 0950\ \delta = 08°\ 11'·1 \pm 0·7'\ (1950)$$
$$CP\ 1919\ \delta = 21°\ 47'·2 \pm 0·7'\ (1950)$$
$$CP\ 1133\ \delta = 16°\ 08'·0 \pm 0·7'\ (1950)$$

The first two values agree very well with the corresponding values obtained at Cambridge with the one-mile radio telescope of $\delta = 08°\ 10' \pm 1'$ and $\delta = 21°\ 47'\ 02'' \pm 10''$, respectively. These values seem to confirm the statement made by Bailey and Mackay[1] that no prominent optical object can be seen within the error field of CP 0950.

A significant difference is apparent between the be-haviour of CP 1919 and that of CP 1133 and CP 0950 as far as the pulse height in a train of pulses is concerned.

CP 1919 usually shows trains of pulses all having almost the same height; the other two show a succession of pulses markedly different from each other. Table 1 shows the percentages of pulses at various normalized peak power intervals for the three sources. $\bar{S}$ is the average peak value of 200 successive pulses for CP 1919 and CP 1133 and of about 400 for CP 0950.

Table 1

Source	$S/\bar{S} < 0.5$	$0.5 < S/\bar{S} < 1$	$1 < S/\bar{S} < 1.5$	$1.5 < S/\bar{S} < 2$	$S/\bar{S} > 2$
CP 1919	2%	68%	27%	3%	0%
CP 1133	55%	25%	13%	3%	4%
CP 0950	52%	16%	10%	6%	16%

At this time of the year the sources are observed respectively in the early morning or in the evening. Interplanetary scintillation should therefore not play a significant part in any of the three sources.

An attempt has also been made to find some sort of correlation between the peak values of the various successive pulses of a given source. For sources CP 1919 and CP 0950, there is no indication of any significant deviation from a sequence of random values, although this does not exclude the presence of a hidden pattern which can be deciphered only by sophisticated filtering.

Source CP 1133, on the other hand, shows some sort of regular pattern—which has appeared in at least three very good and long records—consisting of a fluctuation of the average pulse height with a period which is twice that of the pulse repetition. Our preliminary value for the ratio between the average peak value of "even" pulses and that of "odd" pulses is $1·50 \pm 0·18$. This modulation, if confirmed, would of course support the concept of "something" oscillating both in the fundamental mode and in some higher order mode.

Received May 21, 1968.

[1] Bailey, G. A., and Mackay, C. D., Nature, 218, 129 (1968), (Paper 20).

14. Submillisecond Radio Intensity Variations in Pulsars

by

H. D. CRAFT, jun.,
J. M. COMELLA
F. D. DRAKE

Cornell-Sydney University Astronomy Center,
Arecibo Ionospheric Observatory,
Arecibo, Puerto Rico

PULSARS have so far been observed with time constants and radio frequency bandwidths giving time resolutions of about 1 ms or more[1-4]. We present here preliminary results of observations made at the Arecibo Ionospheric Observatory with equipment adjusted to give a time resolution of 0·1 ms. With this arrangement, pulse structure possessing circular polarization with time scales of the order of 0·2 ms has been seen within single energy from *CP* 0950 and *CP* 1133 (reported at the Conference on Rapidly Pulsating Radio Sources, May 1968). Similar short time structure for *CP* 0950 at a frequency of 2,295 MHz was also reported at the conference by R. D. Ekers and A. T. Moffet. J. H. Taylor has reported observations of occasional circularly polarized pulses made on a lower frequency and with a longer time constant.

Equipment

The observations were made with the 1,000 ft radio telescope. The feed system consisted of the 96 ft radar line feed operating on 430 MHz with orthogonal circular polarizations. Two similar receivers were used, each with a noise temperature of 350° to 400° K. The mixers of the two receivers were driven by the same local oscillator. The IF bandwidths for the two receivers were defined by matched 125 KHz crystal filters centred on 30 MHz. For the sources *CP* 0950 and *CP* 1133, a bandwidth of 125 KHz smears the time resolution by about 0·04 and 0·06 ms, respectively, which is negligible in these observations. The *RC* time constant was 0·1 ms.

energy in the fine structure is not usually a large fraction of the total. These observations contain many cases, as in Fig. 1, in which the observed intensity drops nearly to zero at a time when the mean pulse intensity is large. This suggests the possibility that all of the pulsar radio emission is emitted in bursts of less than 1 ms duration.

The dual-frequency single polarization part of the experiment produced overall pulse structure similar to that in Figs. 1 and 2, that is, a gross pulse structure greatly modulated on a short time scale. Some differences between the pulse structure at the two frequencies were observed, but usually these differences were significantly smaller than those resulting from the dual-polarization experiment.

In the dual polarization observations, some obvious differences in pulse structure exist as can be seen in the illustrations. In interpreting these observations it must be kept in mind that there is a difference in receiver gain between the two channels which has not yet been corrected for. Thus only the changes in polarization within a pulse should be considered at this time. More important is the bandwidth–time constant product of

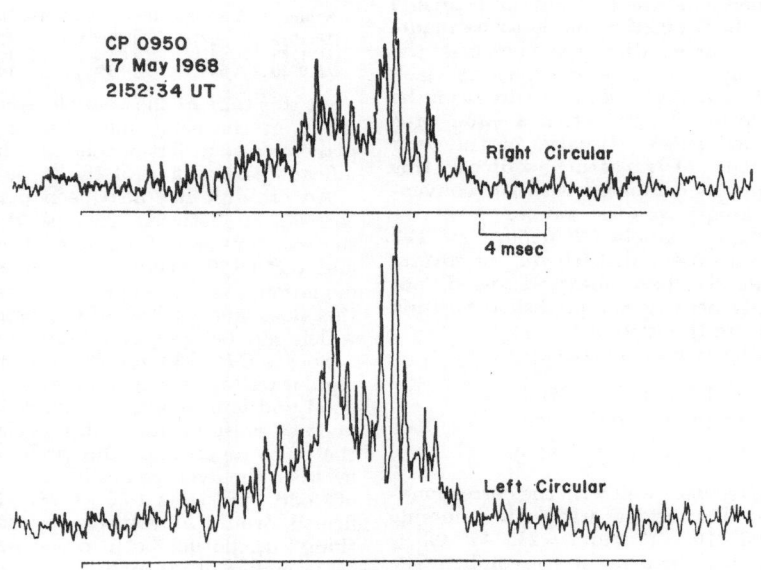

CP 0950
17 May 1968
2152:34 UT

Right Circular

4 msec

Left Circular

Fig. 1. Intensity against time for a single pulse from *CP* 0950 as received on orthogonal circular polarizations. Centre frequency 430 MHz, bandwidth 125 KHz, *RC* time constant 0·1 ms.

The orthogonality of the opposite senses of circular polarization has been periodically checked by the Arecibo planetary radar group. Isolation between orthogonal senses of circular polarization has been measured to be always greater than 10 to 12 dB in the worst conditions of antenna configuration.

In a companion experiment, left circular polarization was observed on two frequencies simultaneously. The receiver IF output was passed through two 100 KHz crystal filters, one centred on 29·5 MHz, the other on 30·5 MHz, permitting simultaneous pulse observations at 429·5 MHz and 430·5 MHz.

Observations

Figs. 1 and 2 show two examples of dual-polarization observations of single pulses of *CP* 0950. Similar results are obtained from *CP* 1133. Short pulses of about 0·2 ms duration can be seen superimposed on what has been called a single pulse component. Although the peak powers in the fine structure far exceed the mean power of the overall pulse over intervals of several ms, the total

25 (allowing a factor of 2 for the fact that $\tau_{\text{eff}} = 2\,\tau_{RC}$). Even in good signal-to-noise conditions, the r.m.s. uncertainty of any point is ± 20 per cent and the uncertainty in percentage polarization is somewhat larger. Thus only intense modulations and polarizations are detectable. Their reality is confirmed, however, by the correlation that exists between independent observations on two polarizations or two frequencies. Yet even with these limitations, there are some significant intrapulse changes in circular polarization. Fig. 2, for example, shows a strong pulse in the left circular channel with no sign of a corresponding pulse in the other channel.

In addition to the intricate pulse structure, Figs. 1 and 2 show another interesting effect, that of apparent reversals of the mean sense of circular polarization across the pulses. Careful inspection, with due regard to the uncertainties in the value of any point, shows that many of the components in both pulses are circularly polarized, and that the sense of this polarization reverses at least four times.

The circular polarization observed is compatible with

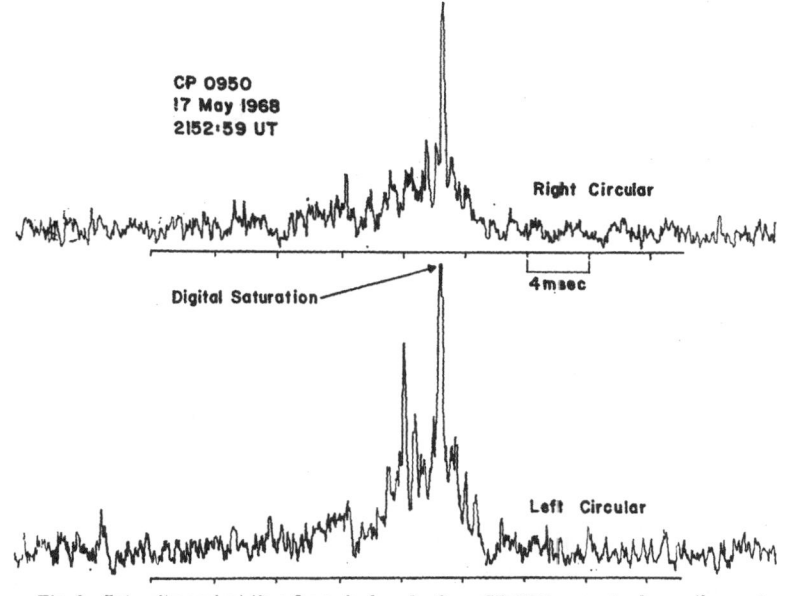

CP 0950
17 May 1968
2152:59 UT

Right Circular

Digital Saturation →

4 msec

Left Circular

Fig. 2. Intensity against time for a single pulse from *CP* 0950 as received on orthogonal circular polarizations. Centre frequency 430 MHz, bandwidth 125 KHz, *RC* time constant 0·1 ms.

the observations of nearly complete linear polarization by Lyne and Smith[3,5] if there are time variations, which is probable, or if only the short time scale components are circularly polarized. The short time scale intensity variations in the orthogonal channels lead to a changing elliptical polarization which goes from one sense of circular polarization to linear to the other circular sense. The lower envelope of the pulse can be either unpolarized or linearly polarized. The occasional circular polarization observed by Taylor is also consistent because a much longer time constant (10 ms) was used in that experiment (Conference on Rapidly Pulsating Radio Sources, New York, May 1968).

Discussion

The intensity variations apparent in Figs. 1 and 2 may be due to scintillations of a single emission source caused by an irregular screen of plasma clouds, or a multiplicity of intense emission regions within the source.

If the variations are caused by scintillations, then the effective scintillation screen scale length must have a size of the order of a Fresnel zone at the screen, and the emitting region must be smaller than or about this size in order to produce the large scintillation index often observed. The radius of a Fresnel zone at a screen near the source of emission is given by $F \approx \sqrt{\lambda R}$ where λ is the radio wavelength and R is the distance from the screen to the source. The velocity of the elements of the screen transverse to the line of sight, v_t, must be such that they move a distance equivalent to about one Fresnel zone during a time interval of the order of the fluctuation time scale, 0·1 ms in this case. Then $v_t = F/\tau$ where v_t is the transverse velocity and τ is the time scale. Table 1 shows a range of values for R, and the related values for F and v_t. The wavelength is 70 cm and τ is assumed to be 0·1 ms. The orbital velocities v_0 of a plasma cloud in orbit at a radius R about an object of one solar mass, making no relativistic corrections, are also calculated. In order that the necessary transverse velocity of

the screen be less than the velocity of light, the diffracting screen must be less than 10^6 km from the source. In addition, F, the size of the emission region, and the scale length of the scintillation screen must be less than 30 km, a remarkably small value. It is interesting to note that the necessary transverse screen velocity is consistent with that of an orbiting plasma cloud at a distance of about 1,000 km from the centre of a one solar mass object, roughly the radius of some plausible white dwarfs. In any case, very high plasma velocities are required in this picture.

Although scintillations can be invoked to explain the rapid intensity variations, they do not directly lead to the observed polarization changes. The large magnetic field and electron density necessary to change the polarization would more likely be found in the source itself than in the surrounding medium.

Turning to the possibility that the overall pulse is composed of emission from many small regions, each would be less than 30 to 60 km in size, and each would have its own polarization. Emission from these small regions would be triggered by a single periodic disturbance propagating outwards from some central object. This model fails to explain the often regular variation of polarization within a single pulse, and the gross pulse structure. The latter feature can be accommodated if the small emitting regions are intense sources within a much larger (6,000 km) structure.

In either model the size of the emitting regions is remarkably small, and radio brightness of these regions is extremely high, very roughly 10^{16} ergs cm^{-2} s^{-1} or about 10^5 times the emittance of the solar surface at all wavelengths. This figure will be reduced by any beaming of the radiation. Nevertheless it indicates the presence of an extraordinary emission mechanism.

The Arecibo Ionospheric Observatory is operated by Cornell University with the support of the Advanced Research Projects Agency through a contract with the US Air Force Office of Scientific Research.

Received June 10, 1968.

Table 1. CHARACTERISTICS OF THE IRREGULAR SCREEN OF PLASMA CLOUDS

R (km)	F (km)	v_t (km s^{-1})	v_0 (km s^{-1})
10	0·084	840	$1 \cdot 1 \times 10^5$
10^2	0·265	2,650	$3 \cdot 8 \times 10^4$
10^3	0·840	8,400	$1 \cdot 1 \times 10^4$
10^4	2·65	$2 \cdot 6 \times 10^4$	$3 \cdot 8 \times 10^3$
10^5	8·40	$8 \cdot 4 \times 10^4$	$1 \cdot 1 \times 10^3$
10^6	26·5	$2 \cdot 6 \times 10^5 \sim c$	$3 \cdot 8 \times 10^2$

[1] Drake, F. D., and Craft, jun., H. D., *Science*, **160**, 758 (1968).
[2] Lyne, A. G., and Rickett, B. J., *Nature*, **218**, 326 (1968), (Paper 10).
[3] Lyne, A. G., and Smith, F. G., *Nature*, **218**, 123 (1968), (Paper 7).
[4] Davies, J. G., Horton, P. W., Lyne, A. G., Rickett, B. J., and Smith, F. G., *Nature*, **217**, 910 (1968), (Paper 6).
[5] Smith, F. G., *Nature*, **218**, 325 (1968), (Paper 47).

15. Measurements of the Pulsed Radio Source *CP* 1919 between 85 and 2,700 MHz

by

B. J. ROBINSON
B. F. C. COOPER
F. F. GARDINER

CSIRO Division of Radiophysics,
Sydney

R. WIELEBINSKI
T. L. LANDECKER

School of Electrical Engineering,
University of Sydney

THE pulsed radio sources[1,2] emit energy almost simultaneously over a wide frequency band[3]. Pulses from the source at 19h 19m 37s, +21°47'02" (epoch 1950) have been received with the Australian 210 ft telescope at frequencies of 85, 150, 630, 1,410 and 2,700 MHz. This report covers measurements of the spectrum of the pulses, the characteristics of the intensity variations at different frequencies, and observations of the structure and polarization of the pulses. A previous communication[4] discussed measurements of the period of the pulses and of the dispersion between 85 and 1,410 MHz.

The pulsed source was first observed at Parkes on March 8, and extensive measurements were carried out between March 13 and 28. Up to four frequencies were received simultaneously; details of the feeds and receivers are given in Table 1. The receivers were used in the total-power mode except for the polarization measurements. The pulses were recorded using pen recorders with a rise time less than 0·1 s, a fast ultraviolet recorder and a R.I.D.L. 400 channel sequential integrating unit driven by a time base derived from a frequency synthesizer. The time base was accurately set to the expected pulse period[4]. Because of the delay in the arrival times of the pulses at the lower frequencies the outputs of three receivers could be multiplexed and integrated in different sections of the 400 word memory. Flux calibrations were made by observing 3C 353 at the start of an observing period and 3C 444 at the conclusion of the observations.

The most surprising characteristic of the pulsed sources is the high degree of constancy of their period[1,4] (for both short and long intervals) combined with large, irregular variations of intensity. The time scale of the intensity

Table 1. CHARACTERISTICS OF RECEIVING SYSTEMS

Frequency (MHz)	Feed	Receiver type	Receiver bandwidth (MHz)
85	Dipole	Transistor	0·4
150	Double dipole	Transistor	1·5
150	Double dipole	Transistor	0·5
150	Double dipole	Transistor	0·15
630	Double dipole	Parametric	10
1,410	Double dipole	Degenerate parametric	10
2,700	Circular horn	Synchronously pumped degenerate parametric	150

variations ranges from seconds to months. We find no correlation in the intensity variations at frequencies between 85 and 1,410 MHz for individual pulses, nor for the average pulse height over intervals of minutes to days.

At 85 and 150 MHz we observe characteristic pulse trains lasting about 1 min, as shown by the chart recordings in Figs. 1 and 2. The pulse amplitude rises progressively for about 30 pulses and then fades back to a low level. There are large pulse-to-pulse variations, particularly at 85 MHz, but the envelope of the pulse train is roughly Gaussian in shape. These trains are seen frequently at 150 MHz and recur quasi-periodically at intervals of 1·5 to 3·5 min. We find no correlation in the arrival time of pulse trains at 85 and 150 MHz, however. Nor is this phenomenon seen at high frequencies; on occasions when the pulses are strong at 630 MHz the intensity remains high for at least 500 pulses. These observations suggest that the rise and fall of the pulse amplitude at the lower frequencies are not generated in the source itself but are impressed by the interstellar medium or a corona around the source. Measurements discussed

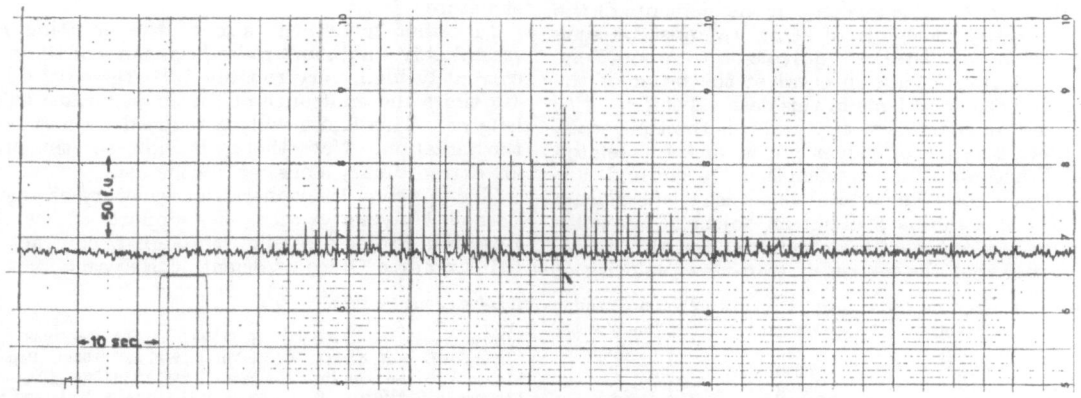

Fig. 1. A train of 150 MHz pulses received at 20h 51m sidereal time on March 8, 1968. The receiver bandwidth was 1·5 MHz. The recorder was slightly underdamped.

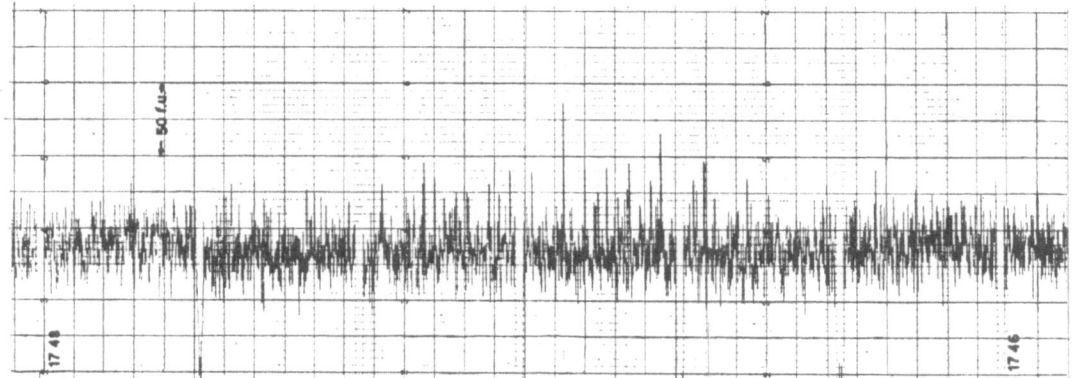

Fig. 2. A train of pulses received at 85 MHz on March 22, 1968. The receiver bandwidth was 0·4 MHz. Sidereal time markers occur at 20 s intervals.

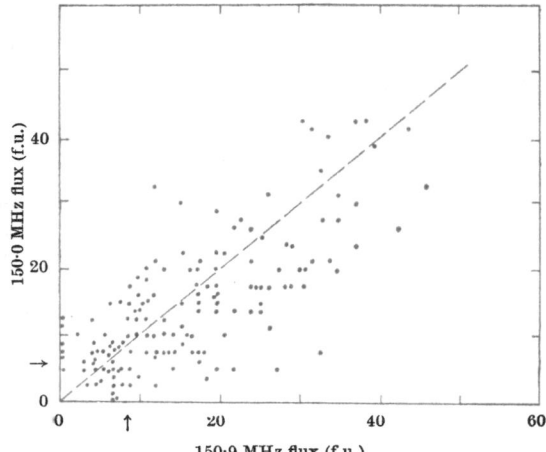

Fig. 3. Scatter diagram showing loss of correlation between pulse amplitudes at frequencies of 150·0 MHz (bandwidth 0·5 MHz) and 150·9 MHz (bandwidth 0·15 MHz). The peak noise levels are indicated by arrows on each axis.

later show that the pulse trains are not produced by a slow rotation of the plane of polarization.

For widely spaced frequencies we have found no correlation between the variations on time scales between seconds and days. For the pulse-to-pulse variations a partial loss of correlation occurs even for a small frequency displacement. Fig. 3 shows the decorrelation between receiver channels centred on 150·0 MHz and 150·9 MHz. If the intensity variations result from interstellar scintillations, the partial correlation between 150·0 and 150·9 MHz gives an estimate of the size of the irregularities[9]. At a distance of, say, 30 pc the irregularity size would be of the order of 10^4 km. If these irregularities were moving with velocities of tens of km s^{-1} the intensity would fluctuate with a time scale of minutes.

Saslaw et al.[5] suggest that the pulses come from a binary neutron star system and that excess activity on one star would produce a correlation between the strengths of alternate pulses. No such effect has been found on the 150 and 630 MHz records. Scott and Collins[6] report no systematic difference between successive pulses at 85 MHz for CP 0950.

On occasions when pulses could be seen on the chart records taken simultaneously at 85, 150, 630 and 1,410 MHz, spectra were determined for individual pulses. When the pulses were weaker at any frequency, average spectra were constructed from 5 min integrations with the 400 word memory.

The range of spectra for individual pulses is shown in Fig. 4. The peak flux varied by a factor of at least 10 at 85 and 150 MHz, and at least 5 at 630 and 1,410 MHz. Peak fluxes of 100×10^{-26} W m^{-2} Hz^{-1} have been observed at the two lower frequencies. For most pulses the peak flux is greatest at 150 MHz, but there is a significant fraction for which the flux is higher at 85 MHz (during 85 MHz pulse trains), and a smaller number for which the 630 MHz flux approaches or exceeds that at 150 MHz. Davies et al.[3] have measured peak fluxes at 408 MHz comparable with the strong pulses we observe at 150 MHz. In general, the spectrum falls rapidly between 630 and 1,410 MHz.

Fig. 4 represents the upper bound to the pulse spectra. At any of the four frequencies the pulses frequently fell below the limit of detection, and the segments of the

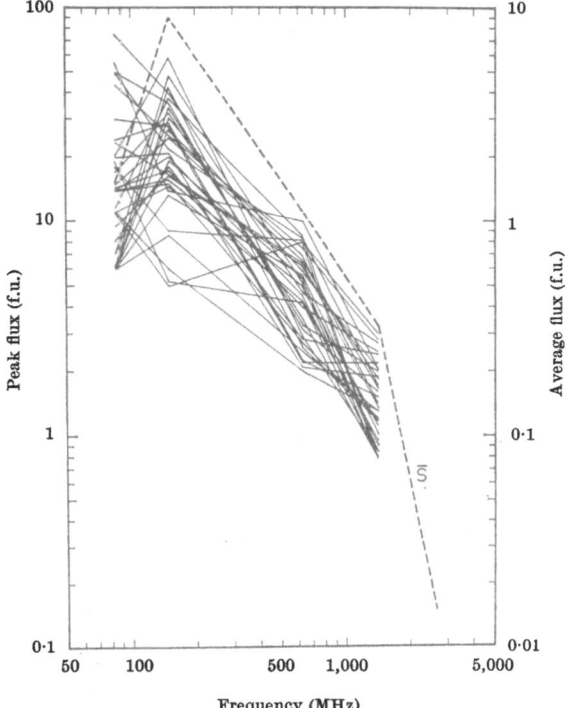

Fig. 4. Spectra of individual pulses between 85 and 1,410 MHz (peak flux scale refers to the continuous lines). The dashed line shows the average pulse height on March 25 integrated over 20 min near 20^h sidereal time (average flux scale). The 2,700 MHz measurement was made on March 28.

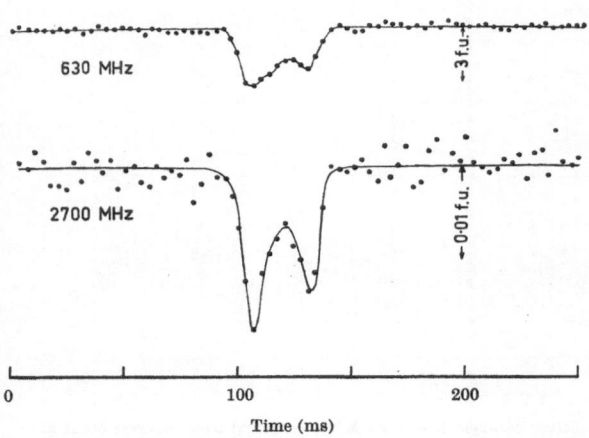

630 MHz

2700 MHz

—3 f.u.

0·01 f.u.

0   100   200

Time (ms)

Fig. 5. Double-pulse structure recorded at 630 and 2,700 MHz. Integration times were 8 min at 630, 20 min at 2,700 MHz.

spectrum became steeper than any shown in the figure. Pulse heights integrated over periods of 5 min gave fluxes an order of magnitude lower. Average spectra obtained from these measurements are shown by the dashed line in Fig. 4. On some days the pulses at 1,410 and 2,700 MHz were below noise even after long integrations. The high-frequency cut-off of the average spectrum can thus be much steeper than in the case shown. Moffet and Ekers[7] also report an extremely low average amplitude at 2,295 MHz. The spectral index between 630 and 2,700 MHz is typically −3·0 to −3·5.

At 630, 1,410 and 2,700 MHz the pulses were always observed to be double, and possibly triple at times. Clear double pulses observed at 630 and 2,700 MHz are shown in Fig. 5. This pulse structure is in essential agreement with that found at 2,295 MHz by Moffet and Ekers[7] and at 408 MHz by Davies et al.[3]. At 2,700 MHz the leading pulse has a rise time of less than 7 ms and a width of 11 ms; the trailing pulse has a similar width and a decay time of about 5 ms. The total pulse duration is about 38 ms.

At 85 and 150 MHz the dispersion across the receiver pass-band results in a blending of the component pulses. The intensities of the components apparently varied

independently as the shape of the blended pulse changed considerably.

Linear polarization measurements were made at 150 and 2,700 MHz. At 150 MHz the receiver was switched at 400 Hz between orthogonal dipole feeds; at 2,700 MHz the outputs from orthogonal polarizations in the horn feed were continuously correlated. In both cases the sum (total power) of the two polarizations was recorded as well as the difference. The average linear polarization at both frequencies (integrated for 5 to 20 min) was found to be less than 10 per cent. A similar null result at 408 MHz has been reported by Lyne and Smith[8] for averages over 1 min.

The polarization of individual pulses at 150 MHz was examined on the chart records. A few per cent of the pulses showed significant linear polarization. Occasional pulses had more than 50 per cent polarization, the degree and orientation being very variable in a time scale of seconds. These rapid variations are probably associated with the multiple-pulse structure of the emission. The relative intensities of the components varied, so one or other could at times dominate the 150 MHz output. The rapid variations of 150 MHz polarization suggest that the individual component pulses are at times highly polarized in different planes. The source CP 1133 shows similar behaviour in the polarization of its discrete components[8]. The 2,700 MHz measurements, in which the pulses from CP 1919 are well resolved, show, however, that any polarization of the components averages out to near zero in a 20 min integration.

We thank D. J. Cooke and D. J. Cole for assistance with the instrumentation. We have benefited from extensive discussion with V. Radhakrishnan and M. M. Komesaroff.

Received May 7, 1968.

[1] Hewish, A., Bell, S. J., Pilkington, J. D. H., Scott, P. F., and Collins, R. A., Nature, 217, 709 (1968), (Paper 1).
[2] Pilkington, J. D. H., Hewish, A., Bell, S. J., and Cole, T. W., Nature, 218, 126 (1968), (Paper 2).
[3] Davies, J. G., Horton, P. W., Lyne, A. G., Rickett, B. J., and Smith, F. G., Nature, 217, 910 (1968), (Paper 6).
[4] Radhakrishnan, V., Komesaroff, M. M., and Cooke, D. J., Nature, 218, 229 (1968), (Paper 17).
[5] Saslaw, W. C., Faulkner, J., and Strittmatter, P. A., Nature, 217, 1222 (1968), (Paper 42).
[6] Scott, P. F., and Collins, R. A., Nature, 218, 230 (1968), (Paper 8).
[7] Moffet, A. T., and Ekers, R. D., Nature, 218, 227 (1968), (Paper 9).
[8] Lyne, A. G., and Smith, F. G., Nature, 218, 124 (1968), (Paper 7).
[9] Budden, K. G., J. Atmos. Terr. Phys., 27, 883 (1965).

16. Spaced Receiver Observations of Pulsed Radio Sources

by

O. B. SLEE
M. M. KOMESAROFF
CSIRO Division of Radiophysics,
Sydney

P. M. McCULLOCH
Department of Physics,
University of Tasmania,
Hobart

THE apparently random fluctuations in the amplitude of pulsed radio sources[1] contrasts strikingly with the strictly periodic nature of the pulses[1,2]. This behaviour suggests that the fluctuations do not originate in the source itself but rather are imposed by the intervening medium. Scheuer[3], however, has argued that the amplitude variations cannot be scintillations produced by either the

general interstellar medium or the interplanetary medium. A spaced receiver experiment which we carried out in May 1968 provides experimental confirmation of Scheuer's conclusion, at least for those fluctuations having quasi-periods of up to several minutes' duration. Our results demonstrate that there is no detectable time displacement between the amplitude variations at the spaced receivers.

This provides direct evidence that variations in pulse amplitude are imposed at or near the source itself.

The observations were made with the 64 m reflector at Parkes, New South Wales, and the 3 km diameter radioheliograph at Culgoora, New South Wales. The baseline from Parkes to Culgoora is orientated about 40° east of north. The hour angle of the baseline is −37° 26′, its declination is 52° 09′ and its length is 325 km. Total power receivers closely aligned in frequency to 80 MHz and with identical rectangular bandpass characteristics of 1 MHz width were used at both sites. The 64 m reflector accepted linear polarization, but the ninety-six reflectors in the radioheliograph were arranged to respond to circular polarization. This scheme eliminated the fading of linearly polarized signals at Culgoora caused by Faraday rotation. Because the radioheliograph has about four times the effective area of the 64 m reflector, the signal/noise ratio at Culgoora on even the highly linearly polarized pulses was generally better than twice the signal/noise ratio at Parkes. The detected signals from the receivers were applied to identical chart recorders, which had response times of ~0·1 s for full scale deflexion and chart speeds of 160 mm min⁻¹.

The pulsars CP 0950, CP 1133 and CP 1919 were each tracked for several hours, the observations extending from May 23 to May 29, 1968. The degree of correlation

amplitudes. A cross-correlation analysis of short sections of the records confirms the impression given by Fig. 1 that the correlation is highest for zero time displacement and is significantly reduced by a relative displacement of one pulse in either direction along the time axis. Similar results were obtained from pulsars CP 0950 and CP 1919.

The results demonstrate clearly that most of the amplitude variations with quasi-periods ranging from a fraction of a second up to several minutes are imposed at or near the pulsed sources. Highly correlated patterns of the kind shown in Fig. 1 cannot be caused by either interplanetary or interstellar scintillations, for the following reasons.

(1) Because of the motion with respect to the Earth of electron irregularities in the solar wind, interplanetary scintillations (if correlated over a baseline of 325 km) should be displaced by ~1 s; our cross-correlation analysis shows that any time displacement, if present at all, must be less than 0·25 s (the pulse period of CP 0950). We thus conclude that any interplanetary scintillations present on our records were not correlated at the spaced receivers; it seems likely that some of the decorrelation evident in Fig. 1 is of this nature.

(2) If it is assumed that the Earth's motion with respect to the interstellar medium is ~30 km s⁻¹, inter-

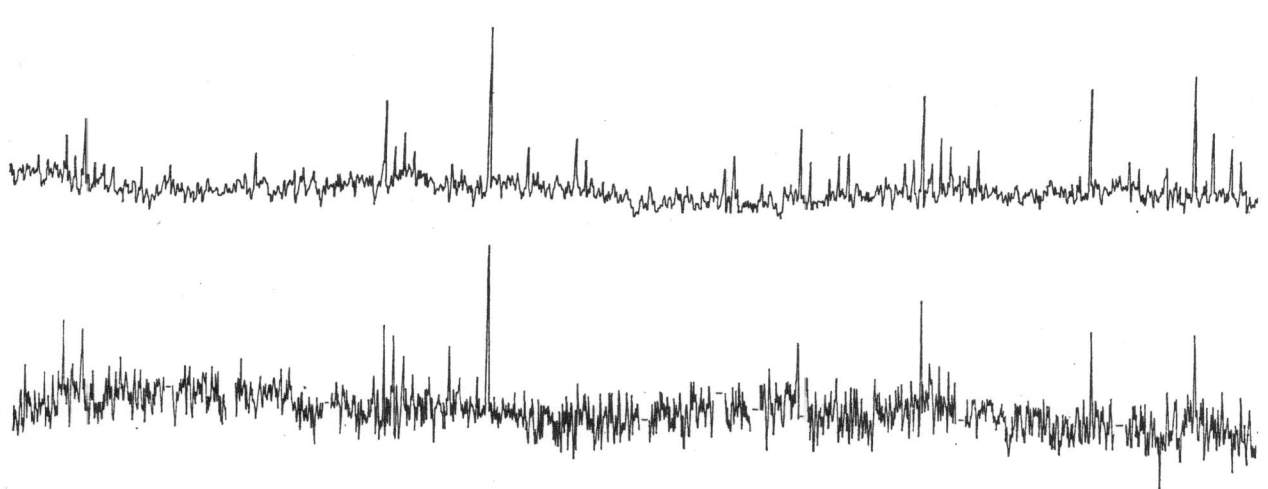

Fig. 1. The upper and lower traces show the pulses from pulsar CP 1133 recorded at Culgoora and Parkes respectively over an interval of 2·5 min on May 28, 1968. The traces have been aligned to within 0·1 s. The lower signal/noise ratio at Parkes is caused by the smaller effective area of the 64 m reflector. The straight sections on the lower trace denote regions of interference or where time marks have been removed.

between the pulse amplitudes at the two receivers varied, not unexpectedly, between wide limits; many of the differences were probably caused by the combined effects of Faraday fading at Parkes and uncorrelated ionospheric scintillations at both sites. Additional complicating factors in the comparison included occasional strong local interference at Parkes and a varying degree of ionospheric refraction, which was sometimes strong enough to shift the pulsar outside the narrow (4′ arc) beam of the heliograph. There were, however, many intervals of 5–10 min when pulses of comparable amplitude were seen consistently at both receivers, indicating that at these times none of these disturbing factors was serious. The results to be discussed here are confined to such occasions.

Short sections of the chart records for pulsar CP 1133 are reproduced in Fig. 1. The charts from the two receivers have been aligned in time to within about 0·1 s. If the high level of noise fluctuation on the Parkes (lower) trace is taken into account, it is clear that most of the pulses arrive at the spaced receivers with similar relative

stellar scintillations would be expected to show time displacements of up to ~10 s over the Parkes–Culgoora baseline. There are no displacements greater than ~0·25 s.

Our experiment has not eliminated the possibility of diffraction by electron irregularities situated close to the pulsed sources and associated with them. In such circumstances the presence of only a very low relative transverse velocity between the source and diffracting region would result in a greatly magnified velocity for the diffraction pattern near the Earth. Much longer baselines would be needed to detect the presence of such fast-moving diffraction patterns.

We thank Mr C. S. Higgins for assistance with the observations at Culgoora.

Received July 2, 1968.

[1] Hewish, A., Bell, S. J., Pilkington, J. D. H., Scott, P. F., and Collins, R. A., *Nature*, 217, 709 (1968), (Paper 1).
[2] Radhakrishnan, V., Komesaroff, M. M., and Cooke, D. G., *Nature*, 218, 229 (1968), (Paper 17).
[3] Scheuer, P. A. G., *Nature*, 218, 920 (1968), (Paper 46).

17. Measurements on the Period of the Pulsating Radio Source at 1919 + 21

by

V. RADHAKRISHNAN,
M. M. KOMESAROFF
D. J. COOKE

Division of Radiophysics,
CSIRO, Sydney, Australia

OBSERVATIONS at several frequencies were recently made at Parkes of the pulsating radio source reported by Hewish et al.[1]. Four receivers centred at 85, 150, 630 and 1,410 MHz were operated simultaneously on the 210-ft telescope during the period March 22–25, 1968. This short communication reports only the measurements made on the arrival time of pulses. Other characteristics of the radiation are dealt with in a separate communication. The highest signal to noise ratio was at 150 MHz and 630 MHz, and it was possible, on the basis of measurements at either of these two frequencies, to establish the period of the pulsations to better than 1 part in 10^7. The combined results for observations during the 4 day period give a value of $1 \cdot 33730109 \pm 0 \cdot 00000007$s.

The output of any one of the receivers could be switched into a 400 channel sequential integrating unit driven by a time base derived from a stable frequency synthesizer. The period T of the time base was adjusted to be close to the period of the pulses corrected for all known motions of the Earth, and the pulses were then integrated for times of the order of 10 min. The time resolution of the system was $T/400 \approx 3 \cdot 343$ ms. The result of the integration was displayed visually on an oscilloscope and plotted on an X–Y plotter. Any discrepancy between the assumed period and the true period manifested itself in a slow drift of the pulse across the channels of the integrator during the 3 h of observation on each day, and also from day to day. The phase of the time base was checked each day by comparing it with time signals transmitted by the Australian Post Office.

The duration of the pulses at any of the observing frequencies is short compared with the interval between them, and, because pulses at different frequencies arrive at different times within the cycle, it was possible to multiplex the outputs of three different receivers into the integrating unit so that all three were sampled by the same time base and integrated in rapid sequence. The arrival times at different frequencies could thus be directly compared, permitting accurate values for the differential delays to be obtained.

Our measurement of the period of the pulsations is based on the assumption that the source position is that given by Ryle and Bailey[2]—namely,

$$\alpha(1950 \cdot 0) = 19^h \ 19^m \ 37^s$$

$$\delta(1950 \cdot 0) = +21° \ 47' \ 02''$$

Corrections for the various motions of the Earth were computed for this position and yielded a value for the true period P_0 with respect to the Sun of $1 \cdot 33730109 \pm 7 \times 10^{-8}$ s. This value is in apparent disagreement with that quoted by Hewish et al. ($1 \cdot 3372795 \pm 2 \times 10^{-6}$ s) by an amount well in excess of the combined errors. The unlikely possibility that the period is a function of frequency is ruled out by our observations, which agree on all four frequencies to within one part in 10^6. Another possibility is that the period itself varies with time. While this alternative is conceivable, the observations of Hewish et al. extending over months gave no hint of any change and in fact showed that the true period was constant to better than 1 part in 10^7. We have examined our data for possible short-term variations in periodicity, but have found none.

The Cambridge observations, except for one 6 h period, were all confined to ± 2 min of transit. While the accuracy of their final value depends on the total span of the observations, the resolution of the ambiguities in the period obtained from the transit observations depends heavily on the single 6 h measurement reported. By making the number of pulses per sidereal day precisely one more than the number we actually find, we arrive at an apparent period $P_0' = 1 \cdot 33728033$, in excellent agreement with the value obtained by Hewish et al. from observations 4 months earlier.

From measurements between any two frequencies we were able to derive a value of the dispersion coefficient $\Delta t/(\lambda_1{}^2 - \lambda_2{}^2)$, where Δt is the difference between pulse arrival times at the corresponding wavelengths λ_1 and λ_2. We find the dispersion coefficient to be independent of frequency, and by combining all our measurements we obtain an average value of $0 \cdot 5770 \pm 0 \cdot 0008$ seconds per square metre. This is in essential agreement with the corresponding value derived from the integrated electron density in the line of sight quoted by Davies et al.[3] as $12 \cdot 55 \pm 0 \cdot 06$ pc cm^{-3}.

The frequency drift rate at a frequency ν calculated from our dispersion coefficient is $\dfrac{\nu^3}{103,700 \pm 150}$ MHz s^{-1}, where ν is expressed in MHz.

The low-frequency receivers were kindly provided by the Electrical Engineering School of the University of Sydney.

We thank W. Miller Goss for the computation of the corrections for the Earth's motion and D. J. Cole for assistance with the instrumentation of the project.

Received April 16, 1968.

[1] Hewish, A., Bell, S. J., Pilkington, J. D. H., Scott, P. F., and Collins, R. A., Nature, 217, 709 (1968), (Paper 1).
[2] Ryle, M., and Bailey, J. A., Nature, 217, 907 (1968), (Paper 19).
[3] Davies, J. G., Horton, P. W., Lyne, A. G., Rickett, B. J., and Smith, F. G., Nature, 217, 910 (1968), (Paper 6).

18. Confirmation of the Parkes Period for *CP* 1919

by
G. ZEISSIG
D. W. RICHARDS
Arecibo Ionospheric Observatory,
Cornell-Sydney University Astronomy Centre,
Arecibo, Puerto Rico

THE measurement of the period of *CP* 1919 by Radha-krishnan, Komesaroff and Cooke[1] disagrees with the value obtained at Cambridge by Hewish *et al.*[2] by about 21 μs, while the combined errors are less than 3 μs (Table 1).

To resolve the discrepancy, we have derived a period for the pulsations of this source from observations made at Arecibo on March 20–21, 1968, close to the dates of the Parkes observations (March 22–25, 1968). The result, corrected to the rest frame of the Sun, is $P_o = 1^s.337\,3017 \pm 0.000\,0005$, agreeing closely with the Parkes number.

The observations were made at a frequency of 111·5 MHz, with IF bandwidth of 100 kHz; the detected signal was filtered with a time constant of 0·01 s. Data were taken at a rate of 250 samples per second by a recorder synchronized to a Varian *R*-20 rubidium standard, which is stable on a 1 h time scale to better than one part in 10^{11}. The system, including the RF section, was checked for stability by transmission of artificial 10 ms wide pulses derived from a Manson oscillator, stable to one part in 10^{10} on a 1 h time scale.

Signal-to-noise of the pulses was sufficient to permit counting of individual pulses. Thus it was possible to obtain, from 41 min of observations on March 20 (10 : 45 : 10–11 : 26 : 46 UT), a mean period with rms deviation small enough to allow prediction of arrival of a pulse 24 h later. This was done by measuring the period of eight pairs of pulses, each pair separated by about 40ᵐ. The deviation from the mean is reduced by $1/\sqrt{8}$ compared with the error in a single such measurement, which is roughly (pulse width)/(number of pulses, *n*, in 40 min). The value of *n* was found by projecting up to 40 min from estimates of apparent period, *P*, over shorter stretches of data.

Six strong pulses that arrived at 10 : 43 : 02–10 : 43 : 11 UT on March 21 were then sufficient to permit extension of the average for *P* during 24 h. The apparent period of $P = 1^s.337\,21459 \pm 14 \times 10^{-8}$ was so derived. The Doppler shift correction, which did not account for acceleration of the Earth, yields the value of P_o given in Table 1.

Table 1. PERIOD OF *CP* 1919 IN REST FRAME OF THE SUN

Observatory	Date of observations	Period	Reference
Cambridge	Dec. 1967–Jan. 1968	1·337 279 5 ± 2 × 10⁻⁶ s	2
Parkes	March 22–25, 1968	1·337 301 13 ± 7 × 10⁻⁸	1
Goldstone	March 16, 1968	1·337 305 ± 20 × 10⁻⁶	3
Arecibo	March 20–21, 1968	1·337 3017 ± 5 × 10⁻⁷	This paper

For further comparison, we have also applied a Doppler correction to the apparent period obtained by Moffet and Ekers[3], estimated errors from data in their paper, and included the derived P_o in Table 1. The agreement among the three latest measurements is very good, further emphasizing the inconsistency with the Cambridge determination. A smooth secular change in P_o seems excluded, for this would have to be of the order of 7 μs per month, on the average, and would have been detected by Hewish *et al*. The possible pulse counting error in the Cambridge work, as suggested by the Parkes group, remains the best explanation.

We thank D. L. Jauncey, F. D. Drake and J. M. Comella for discussions. The Arecibo Ionospheric Observatory is operated by Cornell University with the support of the Advanced Research Projects Agency under a research contract with the Air Force Office of Scientific Research.

Received May 6, 1968.

[1] Radhakrishnan, V., Komesaroff, M. M., and Cooke, D. J., *Nature*, **218**, 229 (1968), (Paper 17).
[2] Hewish, A., Bell, S. J., Pilkington, J. D. H., Scott, P. F., and Collins, B. A., *Nature*, **217**, 709 (1968), (Paper 1).
[3] Moffet, A. T., and Ekers, R. D., *Nature*, **218**, 227 (1968), (Paper 9).

Optical Measurements

19. Optical Identification of the First Neutron Star?

by
M. RYLE
JUDY A. BAILEY
Mullard Radio Astronomy Observatory,
Cavendish Laboratory,
Cambridge

IN a recent communication from this observatory, Hewish *et al.*[1] have described the discovery of a new class of celestial radio source of very small physical size which emits short pulses of radiation with an extremely constant repetition frequency. It was suggested that the radio signals might be associated with the gravitational vibra-tion of a white dwarf or neutron star which could excite repetitive shock disturbances in the stellar atmosphere.

The discovery and detailed study of a neutron star would prove of the greatest importance, both in relation to the properties of super-dense matter and in the mechanism occurring in the gravitational collapse of a star. The possibility that the object discovered by Hewish and his collaborators is indeed a neutron star is thus of great interest, and further investigations by all available methods and over the widest possible range of wavelengths are now required.

The original observations at 81·5 MHz allowed the position to be determined with an accuracy $\Delta\alpha = \pm 3^s$ $\Delta\delta = \pm 30'$. The measurement of the variation of the received repetition frequency caused by the Earth's orbital motion over a period of a month gave an independent value of the declination which was in good agreement with that obtained from the directivity of the aerial system, but the errors were still too large to allow the recognition of any faint optical object which might be associated with the source.

Observations have recently been made at higher frequencies of 408 and 1,407 MHz using the one-mile radio telescope in order (a) to establish a more precise position and (b) to obtain information on the radio spectrum of the pulses. The 81·5 MHz observations have shown that the source is very weak with a flux density in the pulse which varies both from minute to minute and over the months of observation. During most periods the power received from each pulse is $\sim 0\cdot5 \times 10^{-26}$ J m^{-2} Hz^{-1}, so that the mean flux density $S_{81\cdot5} \simeq 0\cdot35 \times 10^{-26}$ W m^{-2} Hz^{-1}. Special recording methods have therefore been used with the one-mile telescope, both to allow the measurement of the small mean flux density and to distinguish the source from the more powerful normal sources which are likely to be found in its neighbourhood. The method makes use of a filtering process based on the characteristic recurrence frequency of the signal both to reject the response due to constant sources and to permit the use of a long effective integration time. In order to avoid the necessity for recording at a high sampling rate, which would be necessary if the filtering were carried out entirely in the computer, the reduction was performed in two stages:

(a) The output of the receiver was switched, using commutating relays, in the time sequence shown in Fig. 1, where the widths of the positive and negative gates are each one-eighth of the total period t, which was made close to the known period of the pulses observed at 81·5 MHz. The commutator was driven from the sidereal crystal clock.

If the commutator is synchronized so that the incoming pulses fall in one or other of the gates, the output from the relays contains a d.c. component proportional to the mean flux density, and integration can be used to increase the signal to noise ratio; the noise level is also reduced relative to that of an unswitched receiver output, while all constant sources produce alternating outputs at the switching frequency which will be rejected by the low-pass filter.

By choosing the period t to differ slightly from that of the incoming pulses (t_0), the output from the filter will alternate with a waveform similar to that shown in Fig. 1, but with a greatly increased period T given by

$$\frac{1}{T} = \frac{1}{t} - \frac{1}{t_0}$$

In practice a value of $t = 1\cdot344328$ sidereal seconds (equal to $1\cdot340654$ solar seconds) was used, giving a value

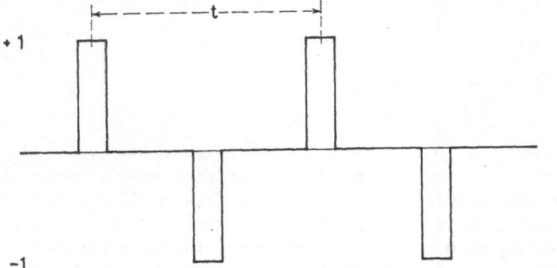

Fig. 1. The time sequence of the commutator.

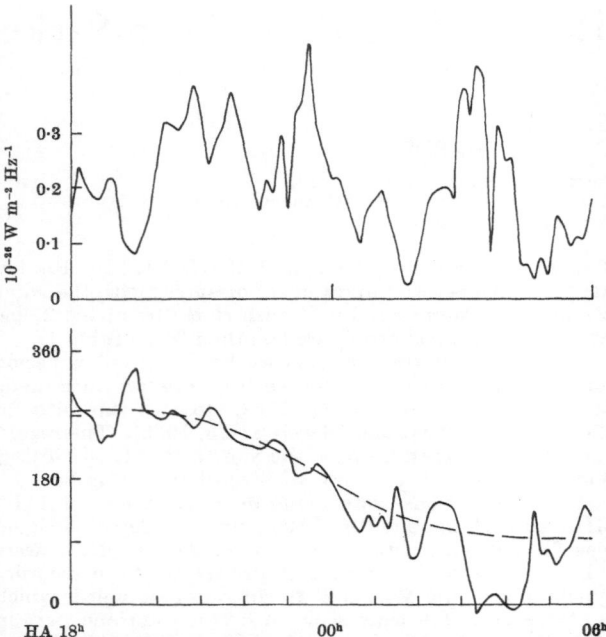

Fig. 2. Record obtained at 408 MHz using an aerial spacing of 900 λ on January 23, 1968. The top trace shows the variation of flux density during the 12^h observing period. The lower trace shows the variation of the radio-frequency phase relative to that of a source at an assumed position close to the actual position. The dashed curve shows the variation of phase computed for a source at the final adopted position.

of $T \sim 520$ s and the signal from the commutator was smoothed with a time constant of 10 s and sampled at intervals of 20 s. This signal was digitized and recorded on punched paper tape in the manner normally used with the one-mile telescope[2].

(b) The reduction in the computer followed that normally employed in the determination of the position of a source[3] except that a convolution operation was first carried out using a function similar to that shown in Fig. 1 but having a periodicity close to T. This operation was necessary because the phase of the incoming pulses and hence of the signal of period T is unknown. The convolution, which corresponded to integration over a period of 30^m, was carried out with displacements of $T/16$. The results were presented as a series of 12^h plots giving the amplitude and radio frequency phase relative to that of a source at an assumed position; the residual variation of the phase then allowed the difference between the true and assumed positions of the source to be determined.

If the values adopted for t and T are sufficiently accurate, the signals will continue, with particular displacements of the convolving function, for the whole 12^h period. If the adopted value of T is incorrect, the signal on any one output will disappear during the run and reappear with another displacement.

One of the output records for the observations of January 23, 1968, with a value of $T = 532\cdot773$ sidereal seconds, which is very close to the optimum value, is shown in Fig. 2; the assumed position was about 4' arc south of the true position, and the dashed line shows the variation of radio-frequency phase appropriate to the final adopted position. From this record and from others reduced with a slightly different value of T (525·0 sidereal seconds) the periodicity of the incoming pulses was derived; when corrected for the orbital velocity of the Earth, a value of $P_0 = 1\cdot337283$ solar seconds was obtained in good agreement with the more accurate value of $1\cdot3372795$ derived by Hewish et al.

The small flux density and the variability of the source make it difficult to obtain an accurate measurement of its position, but the results from three 12^h observations at 408 MHz are presented in Fig. 3. The mean position is

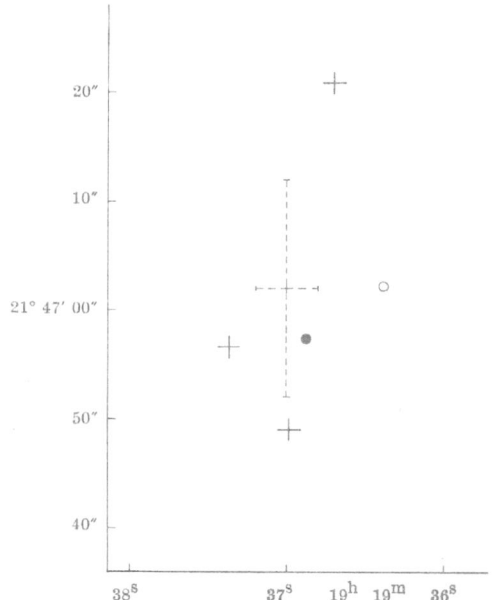

Fig. 3. The positions of the source obtained from successive observations at 408 MHz are indicated by the crosses; the dashed cross represents the mean position and error. The 18^m blue star is indicated by the disk and the faint red object by the open circle.

$$\alpha(1950 \cdot 0) = 19^h \ 19^m \ 37^s \cdot 0 \pm 0^s \cdot 2$$
$$\delta(1950 \cdot 0) = 21° \ 47' \ 02'' \pm 10''$$

A search has been made using an overlay technique[4] to locate the source on the prints of the Palomar Sky Survey, and enlargements of both red and blue prints are shown in Fig. 4. These photographs show that close to the position of the radio source there is an 18^m blue star, with a fainter red object some 8$''$ arc north-west. More precise positions of these objects were therefore obtained by measuring the distances from a number of $AGK2$ stars. The results support the association of the radio source with the blue star the co-ordinates of which are

$$\alpha(950 \cdot 0) = 19^h \ 19^m \ 36^s \cdot 88 \pm 0^s \cdot 1$$
$$\delta(1950 \cdot 0) = 21° \ 46' \ 57'' \cdot 4 \pm 1'' \cdot 5$$

Although this position corresponds to the epoch of the Sky Survey (1950·5) the interpretation of the observed dispersion of the 81·5 MHz signals in terms of the effect of the interstellar medium suggests a distance of 50–100 pc so that any proper motion is unlikely to exceed 3$''$ arc.

The discrepancy in the measured position of only 5$''$ arc suggests strongly that this blue star should be associated with the radio source and further optical observations are now needed. It is possible that measurements of its spectrum and proper motion will provide further evidence to relate it to the radio source. Conclusive evidence would, however, be provided if it were found to exhibit fluctuations of light intensity having the same period as that of the radio emission; the radial oscillations of the star may excite shock fronts in the atmosphere which are responsible for both optical and radio emission. It is possible that electromagnetically accelerated electrons might account for the very high optical surface brightness required without the intense X-ray emission which would accompany thermal excitation; in this case the optical emission might well exhibit the same pulsed characteristics.

Observations at 1,407 MHz, while not having sufficient signal to noise to provide an improved position, have given some evidence on the flux density, and the results have been combined with those at 81·5 and 408 MHz to provide information on the radio spectrum. The spectrum of a given pulse would be of particular interest, and could be determined in principle by correcting for the different times of travel (amounting to some 8 s) derived from the observed dispersion. Such a measurement is not, however, practicable both because of the poor signal to noise ratio in the present observations, and because of the effect of interplanetary scintillation[5] at 81·5 MHz. Instead a mean spectrum has been derived by comparing the average flux densities over the 12^h observing periods with the average flux density observed at 81·5 MHz over a comparable total time during the period December 1967 to February 1968.

The results are shown in Fig. 5. The very high surface brightness (which is probably some 10^5 times greater even than those occurring in QSS) and the absence of a cut-off at low frequencies indicate that the emission cannot be attributed to the synchrotron mechanism[6]; as with the emission from sunspots and from the compact low

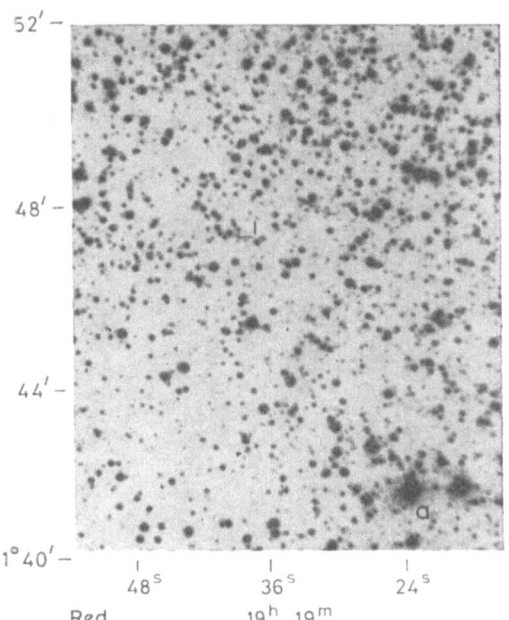

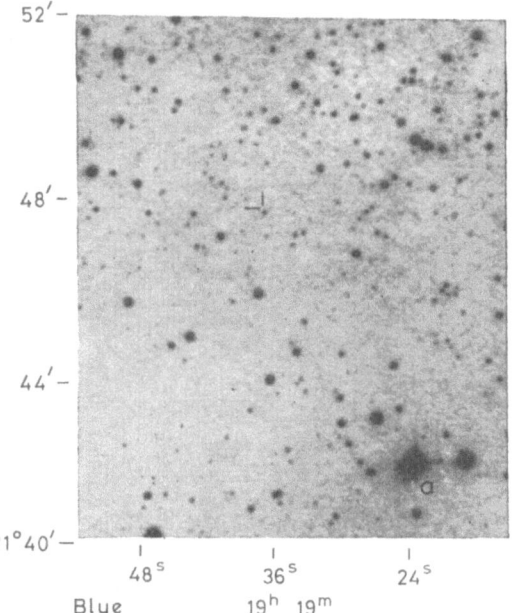

Fig. 4. Enlargements of the red and blue prints of the Palomar Sky Survey showing the radio position. The bright star (a) in the south-west of the area is $AGK2 + 21°$. 1963. (© 1957 National Geographic Society—Palomar Observatory Sky Survey.)

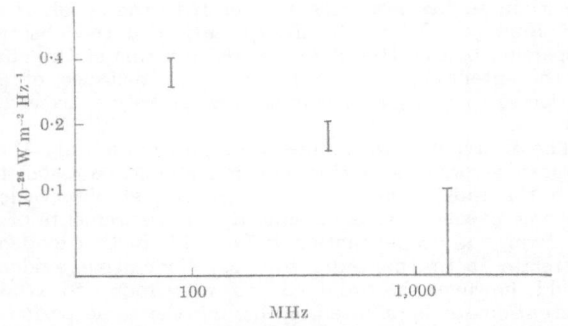

Fig. 5.   The mean flux density of the source at the three observing
frequencies.

frequency source in the Crab Nebula[7], it is more likely
that it is due to coherent electron motion in plasma
oscillations.

Although this new class of object probably only represents about 1 in 5,000 of the total source population at a
given flux density, the remarkable character of their
emission makes it possible to distinguish them for a more
detailed examination of their characteristics.

We thank Drs P. F. Scott and D. M. A. Wilson for help-
ing to assemble the special recording system, and Mr C. D.
Mackay and Dr D. Wills for locating the position of the
source on the Sky Survey prints, and for making accurate
measurements of the position of the blue star.

Received March 4, 1968.

[1] Hewish, A., Bell, S. J., Pilkington, J. D. H., Scott, P. F., and Collins, R. A.,
     Nature, **217**, 709 (1968), (Paper 1).
[2] Elsmore, B., Kenderdine, S., and Ryle, M., *Mon. Not. Roy. Astro. Soc.*, **134**,
     87 (1966).
[3] Parker, E. A., Elsmore, B., and Shakeshaft, J. R., *Nature*, **210**, 22 (1966).
[4] Longair, M. S., *Mon. Not. Roy. Astro. Soc.*, **129**, 419 (1965).
[5] Hewish, A., Scott, P. F., and Wills, D., *Nature*, **203**, 1214 (1964).
[6] Hornby, J. M., and Williams, P. J. S., *Mon. Not. Roy. Astro. Soc.*, **131**, 237
     (1966).
[7] Hewish, A., and Okoye, S. E., *Nature*, **207**, 59 (1965).

20. Accurate Position of a Second Pulsed Radio Source

by

J. A. BAILEY
C. D. MACKAY

Mullard Radio Astronomy Observatory,
Cavendish Laboratory,
University of Cambridge

In a preceding communication, Pilkington *et al.*[1] have
reported the discovery of three more pulsed radio sources
of the type recently described by Hewish *et al.*[2]. One of
these sources, *CP*.0950, is particularly interesting because
its period, 0·25307 s, is considerably shorter than those

of the other three. This source also shows a much more
rapid frequency sweep in the pulse suggesting that the
dispersion introduced by the intervening medium is
smaller; using a reasonable value ($N_e = 0·1$ cm^{-3}) for the
interstellar electron density it seems probable that the

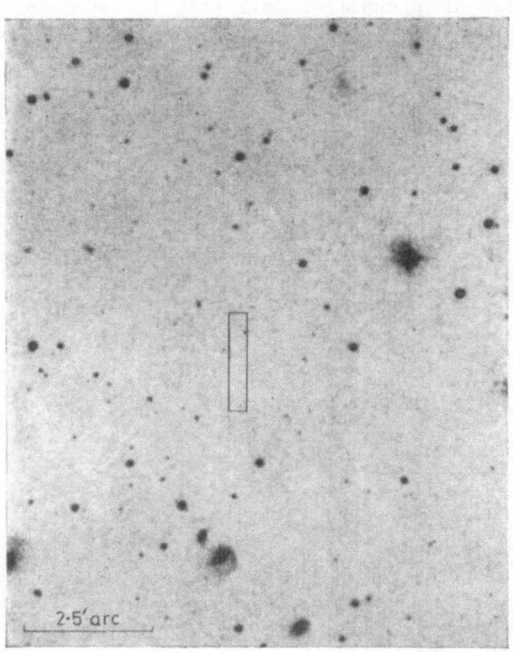

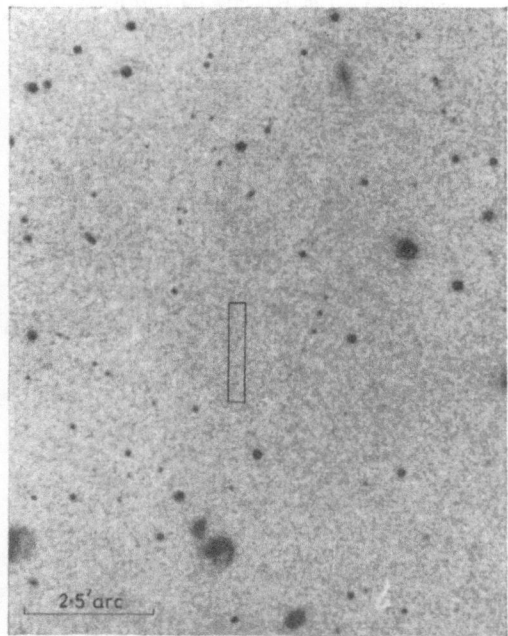

Red   Blue

Fig. 1.   Enlargements of sections of both red and blue prints from the Palomar Sky Survey.   The superimposed rectangle is centred on the
position of *CP*.0950; its size indicates the probable errors of measurement in each co-ordinate.   © 1957 National Geographic Society—Palomar
Observatory Sky Survey.)

distance of $CP.0950$ does not exceed 30 pc. The source lies at a high galactic latitude ($b^{II} = 44°$) so there is therefore a much better possibility of identifying it with an intrinsically faint star than for the case of $CP.1919$ which lies at $b^{II} = 4°$.

Observations have therefore been made with the Cambridge one-mile radio telescope at a frequency of 408 MHz, using the same technique as previously described by Ryle and Bailey[3] for the observation of $CP.1919$. The source shows larger variations of flux density than $CP.1919$ and also lies at a lower declination (where the resolution in declination is worse), but the observations allowed the following position to be obtained:

$$\alpha(1950·0) = 09^h\ 50^m\ 28^s·95 \pm 0^s·7$$

$$\delta(1950·0) = 08°\ 10' \pm 1'$$

A search has been made using an overlay technique[4] to locate the position of the source on prints of the Palomar Sky Survey and enlargements of both red and blue prints are shown in Fig. 1. A finding chart for the area is given in Fig. 2.

These photographs show that just within the error rectangle there is a very faint ($\gtrsim 19^m$) red object barely visible on the blue print. Apart from this object there is no other image brighter than $\simeq 20^m·5$ which can be associated with the radio source, even when account is taken of proper motion which may have occurred since the epoch of the Sky Survey.

New measurements of proper motion and parallax should readily show whether the 19^m object is at a sufficiently small distance to be associated with the radio source.

It has been suggested that the pulsed radio sources may be associated with either white dwarf or neutron stars. It is therefore interesting to compare the apparent magnitude of the observed object with that expected for a white dwarf at a distance of $\simeq 30$ pc.

Allen[5] gives the range of absolute magnitudes of white dwarfs as 10 to 15 (spectral types B_0 to K_0), so that at a distance of 30 pc they would have apparent magnitudes in the range $12·5–17·5$. It is thus clear that, if the 19^m object is in fact associated with the radio pulses, its absolute magnitude is at least $1^m·5$ fainter than that of normal white dwarfs. If measurements of proper motion and parallax reveal that this object cannot be associated with the radio source, the discrepancy is still greater.

The pulsation of this radio source can therefore only be attributed to a harmonic of radial oscillations of a white dwarf—or indeed to any other mechanism associated with a white dwarf—if its absolute magnitude is greater than 16·5, corresponding to a spectral type substantially later than K. These results would also presumably be compatible with a phenomenon associated with a neutron star.

We thank Mr B. Elsmore for his assistance in making the observations.

Received April 3, 1968.

[1] Pilkington, J. D. H., Hewish, A., Bell, S. J., and Cole, T. W., *Nature*, **218**, 126 (1968), (Paper 2).
[2] Hewish, A., Bell, S. J., Pilkington, J. D. H., Scott, P. F., and Collins, R. A., *Nature*, **217**, 709 (1968), (Paper 1).
[3] Ryle, M., and Bailey, J. A., *Nature*, **217**, 907 (1968), (Paper 19).
[4] Longair, M. S., *Mon. Not. Roy. Astro. Soc.*, **129**, 419 (1965).
[5] Allen, C. W., *Astrophysical Quantities* (Athlone Press, 1963).

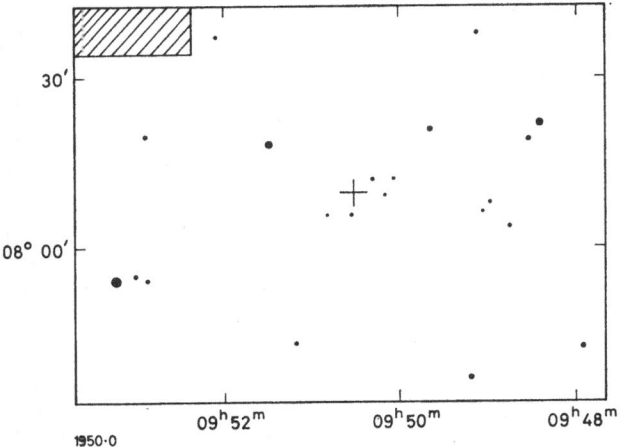

Fig. 2. Finding chart for $CP.0950$ taken from the north-east corner of Sky Survey Print E-233.

21. Accurate Positions of CP 0950 and CP 1133

by

C. D. MACKAY
B. ELSMORE
J. A. BAILEY

Mullard Radio Astronomy Observatory,
Cavendish Laboratory,
University of Cambridge

FURTHER observations with the one mile radio telescope[1] have recently been made in order to improve the accuracy of the positions of the four pulsed radio sources reported by Hewish *et al.*[2].

During the period of observation (May 1–8, 1968), $CP1919$ and $CP0834$ were unfortunately weak, and no improvement could be obtained for the positions already reported[3,5]; good positions were however obtained for $CP0950$ and $CP1133$.

The observations were made at frequencies of 408 MHz and 1407 MHz with interferometer spacings of 750 m and 1500 m. The recording technique was the same as that used earlier in the observation of $CP1919$[3]. By comparing, during a 12 h period, the phase of the interferometer signal with that appropriate to a source at an assumed position, it is possible to derive the true position of the source in the manner described by Elsmore, Kenderdine and Ryle[1].

In the case of $CP0950$, the final position was derived from observations at both frequencies; the increased signal to noise ratio at 408 MHz allowed an improved measurement of the apparent position, but the corrections for ionospheric refraction lead to an uncertainty which is comparable with, but independent of, the uncertainty

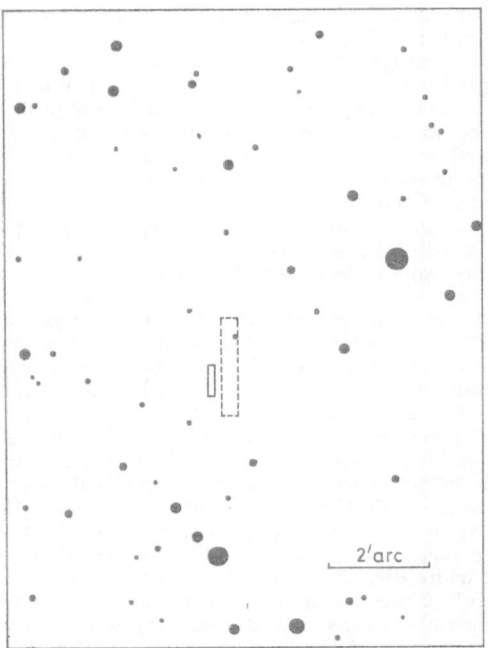

Fig. 1. Finding chart for $CP0950$. The position and errors previously quoted[4] are shown by the dotted rectangle; the solid rectangle indicates the probable errors of the present position. The object in the top right-hand edge of the dotted rectangle is the object suggested earlier as the identification[4].

Table 1. POSITIONS AND PERIODICITIES OF $CP0950$ AND $CP1133$

Source	Position (1950.0)	Period (UT s)
$CP0950$	$09^h\ 50^m\ 30.4^s \pm 0.25^s$ $08°\ 09'\ 47'' \pm 20''$	0.2530646 ± 0.0000002
$CP1133$	$11^h\ 33^m\ 28.1^s \pm 0.6^s$ $16°\ 07'\ 15'' \pm 30''$	1.187911 ± 0.000002

the National Geographic-Palomar Sky Survey. In the case of $CP0950$, the object previously suggested as an identification by Bailey and Mackay[4] can probably be excluded. A sketch of the field of this source is given in Fig. 1; the error rectangle of the original observation is shown dotted; the new rectangle which falls just outside the old one contains no object visible on either red or blue prints. In addition, Luyten (private communication) has compared the Sky Survey plate taken in 1951.01 with another taken in 1963.16, and looked for proper motions of the objects in the field of the source. No object was found which could have moved into the error rectangle in the 17.3 yr since the original Sky Survey plate was taken. There is therefore no optical object that might be associated with $CP0950$ which is brighter than the limiting magnitude of the prints, $m_{pg} \simeq 21$.

In the case of $CP1133$, there is on the red print a very faint image at the print limit near the northern edge of the error rectangle; nothing is visible on the blue print. The apparent magnitude of this object is about 20.5. There is no other optical object within twice the quoted error limits on either print. Finding charts for $CP1133$ are given in Fig. 2, which has been reproduced from the Sky Survey prints.

From the limiting magnitude $m_{pg} \geq 20.5$ which has been found for both $CP0950$ and $CP1133$, an upper limit may be set to the absolute optical luminosity of the object responsible for the radio pulses; for an assumed spectrum similar to that of the Sun

$$L \leq 6 \times 10^{-9}\ L_\odot\ D^2$$

where $L_\odot$ is the luminosity of the Sun, and D is the distance of the object in parsecs. The distance of $CP0950$ has been estimated by Pilkington et al.[5] and by Lyne and Rickett[6]

in the 1407 MHz determination. The position given in Table 1 is a weighted mean of observations at both frequencies.

For $CP1133$, the source was not sufficiently intense at 1407 MHz to allow a good determination, and the position given in Table 1 is that derived from the 408 MHz observations alone.

These positions have been examined on the prints of

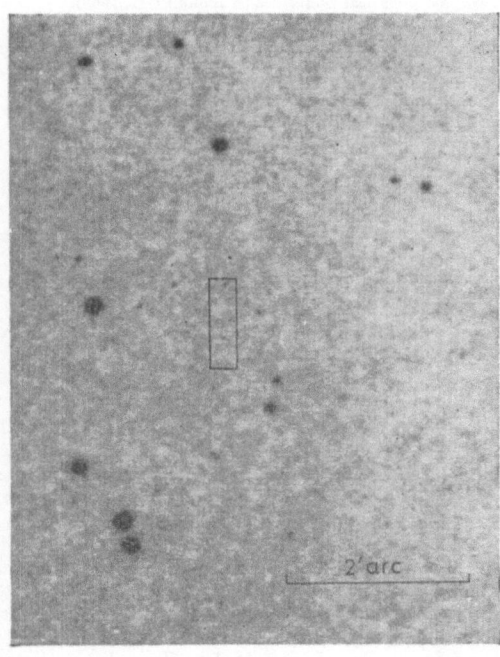

Red

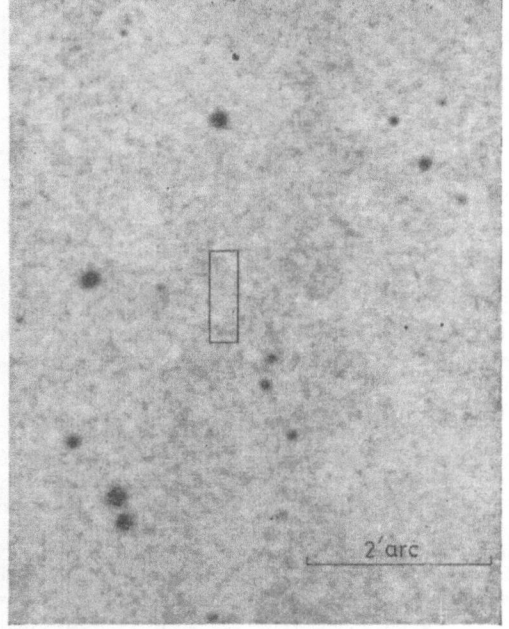

Blue

Fig. 2. Finding charts for $CP1133$, reproduced from the national Geographic-Palomar Sky Survey prints (C). North is to the top, east to the left. The probable errors of the present position are indicated by the error rectangle.

to be ~ 30 pc, assuming the interstellar electron density to be ~ 0.1 cm^{-3}, so that the optical luminosity of $CP0950$ must be $\leq 5.4 \times 10^{-6} L_{\odot}$.

Allen[7] gives the range of absolute magnitudes of white dwarfs of spectral types B_0 to K_0 as $10 \leq M_{pg} \leq 15$; the absolute magnitude of $CP0950$, on the other hand, must be $M_{pg} > 18$. This result places an important limit on possible models of pulsating radio sources.

Arp (private communication) has recently taken photographs with the 200 inch Hale telescope of the field of $CP1919$, which reach down to a limiting magnitude $m_{pg} \simeq 23$. The improved accuracy of the present observations justifies similar observations of $CP0950$ and $CP1133$; the results might either reveal the sources or set still lower limits to their optical luminosity.

In the course of these observations, improved periodi-

cities were obtained for both sources in the manner described by Ryle and Bailey[3]. The new values are given in Table 1.

We thank Professor Sir Martin Ryle for helpful discussions, and Mr T. W. Cole for assisting in the computation of the periodicities.

Received June 24, 1968.

[1] Elsmore, B., Kenderdine, S., and Ryle, M., *Mon. Not. Roy. Astro. Soc.*, **134**, 87 (1966).
[2] Hewish, A., Bell, S. J., Pilkington, J. D. H., Scott, P. F., and Collins, R. A., *Nature*, **217**, 709 (1968), (Paper 1).
[3] Ryle, M., and Bailey, J. A., *Nature*, **217**, 907 (1968), (Paper 19).
[4] Bailey, J. A., and Mackay, C. D., *Nature*, **218**, 129 (1968), (Paper 20).
[5] Pilkington, J. D. H., Hewish, A., Bell, S. J., and Cole, T. W., *Nature*, **218**, 126 (1968), (Paper 2).
[6] Lyne, A. G., and Rickett, B. J., *Nature*, **218**, 326 (1968), (Paper 10).
[7] Allen, C. W., *Astrophysical Quantities* (Athlone Press, 1963).

22. Positions of Four Pulsars

by

A. J. TURTLE
A. E. VAUGHAN

The Chatterton Astrophysics Department,
School of Physics,
University of Sydney

ACCURATE positions of four pulsars[1] have been measured with the one mile cross type radio telescope at the Molonglo Radio Observatory. This instrument[2], which operates at a frequency of 408 MHz, forms eleven beams simultaneously, each 2.7' in right ascension by 2.8' in declination and separated by 1.4' in the north–south direction. Simultaneously with this the east–west arm can be used alone, giving a fan beam in the meridian plane 2.0' by 4.2°. Similarly, there are eleven simultaneous fan beams available from the north–south arm; these are elongated in the east–west direction. Results from all these systems have been combined here. $CP1919$ is, however, right on the northern limit of the east–west system.

The telescope is intended for transit observations only, so any source is in the beam of the east–west arm of the cross for about 10s each day and in the fan beams of the north–south arm for 10 min. This is a disadvantage, as the pulsars are clearly detectable only on about one day in five. But on the occasions when they are active at transit their positions can be measured accurately. Fig. 1 shows the output of the eleven cross beams for May 22, 1968, when $CP0950$ was active and for May 21 when it was undetectable. Individual pulses cannot be seen because the receiver outputs are smoothed with the time constant of 3s normally used in this form of recording.

Right ascensions were determined from the centroid of the times of the pulses, weighted by their intensity, as observed using the fan beam of the east–west arm and a receiver with a short time constant (~ 0.1s). By combining the results of several days' observations the effects of the fluctuations in pulse intensity during transit are reduced. The pulsar $CP0950$, for which the pulse amplitude varies most rapidly, was fortunately observed frequently. Declinations were obtained in two ways. The first method used the value, averaged over many pulses, of the relative response to individual pulses as observed with a short time constant on adjacent north–south arm fan beams. The same principle has been used by Grueff et al.[3]. The linearity of the recording system was calibrated by means of artificial pulses of similar duration. Second, the smoothed cross records similar to

those illustrated in Fig. 1 were used. The area of the deflexion on a given trace is a measure of the total energy received by that beam from the pulsar and its declination can be derived from the ratio of these areas on adjacent beams. Both right ascensions and declinations were calibrated by observations of constant sources with accurately known positions.

Table 1 gives the 1950.0 positions equivalent to the mean position in the observing epoch May–June 1968. Right ascensions are given for the four sources and declinations for three of them. Despite a search with the cross between $+06° 40'$ and $+05° 33'$ the declination of $CP0834$ has not been measurable. We may have been observing with the wrong set of eleven beams on the rare occasions when the pulsar was active, or it may have been confused by the response to an uncatalogued weak constant source which is within 1.1s of arc of the pulsar right ascension and 2' of the declination, and which is given in a summary by Maran and Cameron[4]. This source has a flux density of 1.2×10^{-26} W m^{-2} Hz^{-1} and its position is included in Table 1. The probability of this source falling so close to the proposed pulsar position is small and it may be that there is some association between them. There is no constant source stronger than 0.1×10^{-26} W m^{-2} Hz^{-1} within 5' of $CP0950$ and $CP1133$.

The Palomar Sky Survey plates have been searched at these positions with the following results. The technique of using transparent overlays and measuring to the nearest catalogued star provides an accuracy of a few seconds of arc.

$CP0950$. No objects visible at this position.

$CP1133$. There is a faint red image at 11h 33m 27.9s, 16° 07' 37", and another at 11h 33m 26.4s, 16° 07' 19";

Table 1. PULSAR POSITIONS

	R.A. (1950.0)				δ (1950.0)		
	h	m	s	°	°	'	"
$CP0950$	09	50	30.76 ± 0.15	$+08$	09	48 $\pm$ 5	
$CP1133$	11	33	26.90 ± 0.15	$+16$	07	35 $\pm$ 15	
$CP1919$	19	19	37.2 ± 0.3	$+21$	47	14 $\pm$ 15	
$CP0834$	08	34	26.6 ± 0.5				
Constant source near							
$CP0834$	08	34	25.5 ± 0.5	$+06$	08	44 $\pm$ 10	

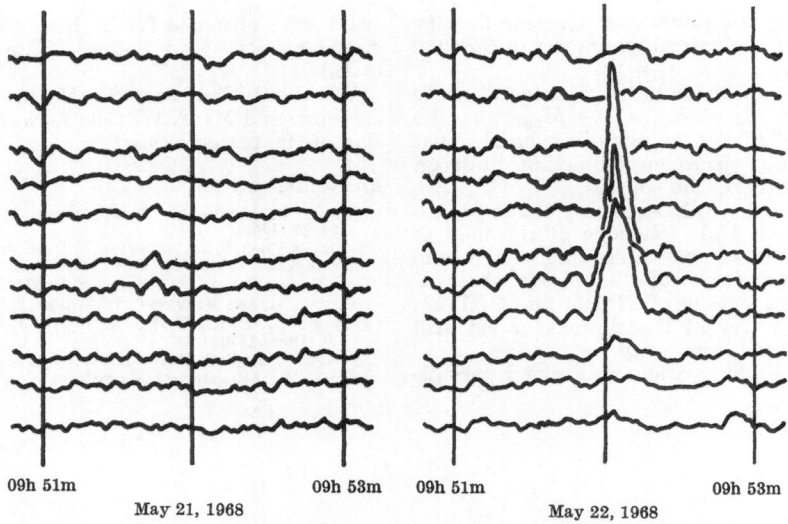

Fig. 1. Recording of the output of eleven cross beams during transit of *CP* 0950 on May 21 and 22, 1968, smoothed with a 3s time constant. The indicated times do not give right ascension directly.

Table 2. POSITIONS OF OPTICAL OBJECTS NEAR THE PROPOSED POSITION OF *CP* 0834

	R.A. (1950·0)			δ (1950·0)			Approx. magnitude
	h	m	s		′	″	
A	08	34	26·31	+06	10	27	15
B	08	34	27·03	+06	10	23	14
C	08	34	26·75	+06	08	46	19
D	08	34	26·83	+06	08	30	11
E	08	34	26·85	+06	06	25	9·0

the latter may be a galaxy. They are both outside the error rectangle, but within it and between the two there is a fainter image barely detectable on the red plate only. Its position is 11h 33m 27·0s, 16° 07′ 26″.

CP 1919. The present radio position is farther from the possible identification (19h 19m 36·88s ± 0·1s, +21° 46′ 57·4″ ± 1·5″) assigned by Ryle and Bailey[5] than the Cambridge radio position (19h 19m 37·0s ± 0·2s, 21° 47′ 02″ ± 10″) but is still compatible with it. On the red plate there is a fainter object to the north at 19h 19m 37·0s, 21° 47′ 26″, which is also just in the error rectangle; no other objects are visible in the error rectangle.

CP 0834. The lack of a precise declination prevents an attempt at identification but there are a number of bright objects with the same right ascension and within a few minutes of arc of the declination of the constant source. Their positions and estimated magnitudes are listed in Table 2. There is no object within the error rectangle around the constant source.

In no case is there a convincing optical identification

for a pulsar down to the limit of the Sky Survey plates. *CP* 0950, which is presumably the nearest of these pulsars, has no optical feature at all. This suggests that optically these sources have a very low intrinsic continuous luminosity.

A system to produce more than one east–west beam will soon be installed. More accurate values of the right ascensions of the pulsars can then be obtained from the ratio of individual pulse amplitudes on each beam averaged over many pulses. By comparison with the position of a convenient constant source it is estimated that changes in the right ascension of a pulsar during the year of less than 1″ may be detected. A useful upper limit to the parallax can hence be established.

We thank A. G. Little and M. J. L. Kesteven for their assistance, and R. W. Hunstead for advice on optical positions. This work was supported by grants from the Australian Research Grants Committee and the US National Science Foundation. One of us (A. E. V.) holds a Commonwealth postgraduate research studentship.

Received July 12, 1968.

[1] Pilkington, J. D. H., Hewish, A., Bell, S. J., and Cole, T. W., *Nature,* **218**, 126 (1968), (Paper 2).
[2] Mills, B. Y., Aitchison, R. E., Little, A. G., and McAdam, W. B., *Proc. IRE Austral.,* **24**, 156 (1963).
[3] Grueff, G., Roffi, G., and Vigotti, M., *Nature,* **218**, 1037 (1968), (Paper 13).
[4] Maran, S. P., and Cameron, A. G. W., *Kitt Peak National Observatory Contribution No. 327* (1968).
[5] Ryle, M., and Bailey, J. A., *Nature,* **217**, 907 (1968), (Paper 19).

23. Time Resolved Photometry in Two Pulsed Radio Source Fields

by

J. BORGMAN
J. KOORNNEEF

Kapteyn Observatory,
Roden, The Netherlands

FOUR pulsed radio sources have so far been discovered[1]. All of them have short duration pulses and periods between 0·25 and 1·34 s.

Soon after the announcement of the discovery of the

first source (*CP* 1919), Ryle and Bailey[2] suggested that this source could be identified with an 18m blue star. In an attempt to confirm the identification, Duthie *et al.*[3] tried to observe optical pulsations of the source. Their detection technique is potentially capable of detecting light pulsations of 0.1^m in a mode similar to the radio observations. The result was negative. A similar conclusion was reached by Bingham[4].

Bailey and Mackay[5] have searched the Palomar Sky Survey at the position of the source $CP\ 0950$; they could only find a very faint red object within the position error rectangle. $CP\ 1919$ has tentatively been identified with a faint blue object, so it seems likely that at least one of these identifications is incorrect.

The positions of the other two sources, $CP\ 0834$ and $CP\ 1133$, are considerably less certain[6]. Although the attempts to find optical pulsations of the source $CP\ 1919$ had negative results, we have tried to find within the position error rectangle objects with light variations in a mode similar to the observed radio pulsations.

Our detection technique is based on photographs of the fields taken with the 24 in. telescope of the Kapteyn Observatory. This telescope has been equipped at the Cassegrain focus with a camera attachment consisting of a large field lens and a 35 mm camera. The system becomes effectively a 24 in. $F/3.3$ telescope, capable of photographing 18^m stars on Kodak 'Tri-X' film in 5 min. The unvignetted field is approximately 15' of arc diameter. In order to obtain time resolved photometry over the entire field the camera was rocked at the published radio pulse frequency by a rotating eccentric wheel driven by a synchronous motor. As a result, "normal" star images are trailed; the objects which we searched for should have shown up as circular images. Several exposures were made to cover the entire position error rectangles.

No optical pulsations were detected. If there were stars within the position error rectangle brighter than seventeenth photographic magnitude and showing pulsations of the off–on type (with the on-portion shorter than 30 per cent of the period), they would have been found.

This means that either the optical counterparts of the radio sources must be fainter than seventeenth photographic magnitude, or that they do not show light pulsations of the same frequency as the radio objects, in which case they may have any brightness. The search for optical identifications in the cases of $CP\ 1919$ and $CP\ 0950$, however, indicates that the optical counterparts cannot be brighter than eighteenth and twentieth photographic magnitude, respectively.

Combined with the information that $CP\ 0834$ and $CP\ 1133$ are within a distance of 100 pc, we must conclude that their brightness does not exceed twelfth absolute photographic magnitude; probably they are considerably fainter.

The possible optical counterparts of $CP\ 1919$ and $CP\ 0950$ must be fainter than thirteenth and fifteenth absolute photographic magnitude, respectively. If the tentative identifications[2,5] with a blue and red star turn out to be correct, the optical counterparts fall in the class of the faintest known white dwarfs.

We thank our colleagues Bosma and Wesselius for their interest in this investigation.

Received April 29, 1968.

[1] Pilkington, J. D. H., Hewish, A., Bell, S. J., and Cole, T. W., *Nature*, 218, 126 (1968), (Paper 2).
[2] Ryle, M., and Bailey, J. A., *Nature*, 217, 907 (1968), (Paper 19).
[3] Duthie, J. G., Sturch, C., and Hafner, E. M., *Science*, 160, 415 (1968).
[4] *I.A.U. Circular No. 2066.*
[5] Bailey, J. A., and Mackay, C. D., *Nature*, 218, 129 (1968), (Paper 20).
[6] *I.A.U. Circular No. 2064.*

24. New York Conference on Pulsars

May 21

NOT long ago it was the quasar, and now it is the pulsar, the radio properties of which burst on us without any corresponding optical information. At this conference there was a wealth of information about the structure of the radio pulses, many ideas on their generation, on the scintillation phenomena which so evidently modulate their strength, and on the nature of the accurate periodicity of the pulses, but there was still very little about the possible identification of any of the four known pulsars with visible objects.

The one new piece of optical information fell as a bombshell at the conference at the Goddard Institute of Space Studies; it was reported both from Kitt Peak and Lick Observatories that the light from Ryle's star, provisionally identified with $CP\ 1919$, was varying by about 4 per cent with a period twice the interval between the radio pulses. If this is true, it makes the identification more certain and the theories more difficult, since there is no evidence at all for the differences one might then expect between alternate radio pulses.

The confusion between various observations was worsened by the dramatic arrival of Dr Cudaback from Lick Observatory, carrying batches of data showing that the light output of $CP\ 1919$ varied by more than 15 per cent with a wide variety of periods, not locked accurately to the radio period but including the double period and its harmonics. This behaviour is clearly not the same as the radio behaviour, and at this early stage it could not be taken into account in the theoretical discussions.

The radio evidence confirmed and enlarged the picture which has emerged from recent papers in *Nature*. New observations at Arecibo and Greenbank have shown that the polarization of radio pulses is typically elliptical rather than plane, and that it varies over the whole range from circular to plane. An extra pulse has been found in Jodrell Bank recordings of $CP\ 0950$; this pulse occurs 100 ms earlier than the main pulse, and contains 1.5 per cent of the energy of the main pulse. Observations of the radio spectrum have been made down to a wavelength of 13 cm for all four sources by the California Institute of Technology, and down to 11 cm for two sources by Jodrell Bank. The spectrum falls steeply for $CP\ 0834$ and $CP\ 1919$, and the radio emission mechanism must explain this very rapid cutoff.

The scintillations of these radio sources, giving the variations shown in Jodrell Bank recordings from a wavelength of 2 m to 30 cm, have the characteristics of focusing at the short wavelengths and a more random character at long wavelengths. Further, there is fine frequency structure in the spectrum which is troublesome for further observation of the Faraday effect in the galactic magnetic field, but which is now clearly attributable to scintillation. Professor F. G. Smith showed that the scintillation occurs in a physically thin shell of ionized gas round the star, with radius perhaps 10^{10} km with irregularities 10^4 km across containing an electron density of about 10^3 cm^{-3}.

The very accurate time-keeping properties of the pulsars suggest that they are either massive oscillating stars or binary systems. A single star would be a white dwarf, oscillating in an overtone mode, and driven by nuclear burning. Binary systems are subject to enormous gravitational forces, and they will be unstable if the orbital period is made short enough. Any single star rotating at the observed pulse period would only be stable if it were as small as a neutron star, but J. P. Ostriker showed that there is a wide range of white dwarf models which rotate fast enough to distort them into ellipsoids. Suggested emission mechanisms varied from a permanent flare, emitting like a lighthouse, to versions of shock waves converting mechanical energy into radiation as they encounter a magnetic field. The shape of the individual spikes and the pulse envelopes fit well with a source distributed over the surface of a white dwarf, according to calculations by B. H. Bland of Manchester.

The conference was prepared to accept interpretations based on white dwarfs, but there is still room for many questions on pulsation theory and the emission mechanism. The observational question about the optical emission also remains open; it may be settled very soon, since the whole of the astronomical world is so excited about the possibility of light pulses that many telescopes will be used on the pulsars during the next few nights.

F. G. SMITH

25. Search for Optical Flashes from the Radio Source *CP* 1919

by

J. V. JELLEY
AERE, Harwell

R. V. WILLSTROP
University of Cambridge Observatories

THE four rapidly pulsating radio sources discovered by Hewish *et al.*[1] have been tentatively identified with neutron stars or white dwarfs, and their optical identification is therefore of great interest.

An accurate position of the source *CP* 1919 has been determined by Ryle and Bailey[2]. Their position and error rectangle,

$$\alpha\ (1950.0) = 19^h\ 19^m\ 37 \cdot 0^s \pm 0 \cdot 2^s$$
$$\delta\ (1950.0) = +21°\ 47'\ 02'' \pm 10''$$

include a yellow object of magnitude about $B = 19$. Just outside the error rectangle there is a red object about 8″ arc north preceding the yellow object. The nearest comparable objects are 25″ arc following the yellow object. The radio measurements of position therefore favour strongly the identification of the pulsating source *CP* 1919 with the yellow object. If this is not the source of the radio signals, the source must be fainter than $B = 21$, the limit of the Palomar Sky Survey, or it must lie outside the error limits. This is believed to be very unlikely[3].

We have looked for fluctuations in the brightness of all objects in two small areas of the sky centred on the yellow object, at the same periodicity as the radio fluctuations. If any such fluctuations were observed it would be possible to deduce the time-averaged brightness of the object emitting the flashes, and if this were brighter than $B = 21$ the object would be expected to appear on the Palomar Sky Survey prints. Hence a positive optical identification might be made. Conversely, if no fluctuations could be detected it would be possible, given sufficiently sensitive equipment, to set an upper limit to the fraction of the light of the yellow object which might be emitted in flashes. If other optical evidence confirms the identification of the radio source with the yellow object, the upper limit to the irregularity of its visible light may help in the determination of its nature and physical properties.

No fluctuations in brightness with a period of 1·3372 s have been detected. On the morning of April 5, 1968, an area 10″ × 10″ arc, centred on the yellow object and excluding the red object, was examined, and on the following morning an area 30″ × 30″ arc, again centred on the yellow object but this time including the red object, was examined. The equipment is described here and there follows an estimate of the limit of sensitivity of the equipment.

The Cambridge Observatories 92 cm reflecting telescope was used, with a photometer at the prime focus. The photometer contained an EMI 9502 photomultiplier, with Sb-Cs photocathode. A Fabry lens of glass, 4 mm thick at the centre, was used, and the window of the photomultiplier housing was a quartz disk 2 mm thick. A small gap between the quartz window and the photomultiplier was filled with a disk of 'Perspex' 1·6 mm thick. Reflexion losses at some of the surfaces were avoided by "cementing" the quartz, 'Perspex' and photomultiplier together with a few drops of dibutyl phthalate. The band pass of the photometer was therefore approximately 3500–6500 Å.

The telescope was guided on the 6·5m star BD + 21° 3740 (= AGK2 + 22° 1939) at $\alpha(1950) = 19^h\ 18^m\ 46^s$, $\delta(1950) = +22°\ 06'\ 14''$. The photometer diaphragm was then set, first visually and then by noting the response of the photometer, on the 9·2m star BD + 21° 3746 (= AGK2 + 21° 1963) at $\alpha(1950) = 19^h\ 19^m\ 23 \cdot 4^s$, $\delta(1950) = +21°\ 41'\ 42''$. The whole photometer was then offset, using micrometer screws, through calculated distances to reach the yellow object. The micrometers could be read to 0·01 mm directly, and the smaller of the diaphragms used was 0·20 mm square. The accuracy of the movement of the photometer was checked by observing a number of stars in the open cluster Praesepe while using the smaller diaphragm.

The sensitivity of the photometer was determined by observing stars in Praesepe and a star 0·8′ arc north and 0·8′ arc following BD + 21° 3746. This star has been estimated by Liller (unpublished work) to have $B = 13·6$. The transparency of the sky in Great Britain cannot be relied on to remain uniform or constant for times of several hours, so we have used Liller's magnitude in estimating the limiting sensitivity of our equipment. The response of the photometer to stars in Praesepe, which have magnitudes determined photoelectrically by Johnson[4], did not suggest that Liller's estimate was in error by more than 0·1m or 0·2m.

The output of the photomultiplier was recorded by using pulse-counting techniques. It is worth remarking that the time resolution provided by this technique, which is not easily obtained by other astronomical photometric techniques, is vital to this work.

The photoelectron pulses from the photometer were fed to a Laben analyser operating in the multiscaler mode. The pulses were fed into 512 counting channels which were switched in a serial manner, so that each channel counted for approximately 2·6 ms per cycle. A new cycle was initiated every 1·337 s, very shortly after the end of the counting in the last channel. The overall deadtime was approximately 0·2 per cent of the full period.

In this initial experiment it was assumed that the heliocentric period given by Hewish et al.[1] was correct, and that the light flashes might last 16 ms[1] or perhaps 35 ms[5]. A recent paper by Radhakrishnan, Komesaroff and Cooke[6] has suggested that the heliocentric period is 1·33730113 s; the implications of this are discussed later.

The essential features of the equipment are shown in Fig. 1. The efficacy of the method depends solely on the precision of the driving oscillator and scaling circuits, because errors in timing are cumulative throughout a run. Three oscillators were used, as follows. The basic standard of frequency, A, Fig. 1, was derived from the 200 kHz carrier of the BBC Droitwich transmitter, available each night until 0200 h BST. During daytime and early evening the frequency of a local standard 5 MHz crystal oscillator C was adjusted as closely as possible by direct comparison with A. This oven controlled reference was then used to cover the observing period, approximately 0200 h to 0400 h UT (0300 to 0500 BST). The analyser itself was run from a lower grade 1 MHz crystal, B, the frequency of which was continuously monitored against C, and adjusted manually throughout the runs.

A six-digit dividing circuit was used to derive the basic cycling period. This unit was set to give a period of 1·3371880 mean solar seconds. This value, used throughout the observations, was the closest that could be achieved to the period 1·3371860 s obtained after correcting for the Earth's orbital motion of −20·958 km s[-1] in the line of sight for April 2d 12h, 1968. A second dividing circuit provided pulses spaced by 2·6 ms to drive the channel advance circuits. The errors in the overall timing accuracy are listed in Table 1.

The overall frequency performance of the oscillators was thus adequate to maintain synchronism for a 40 min run to a precision of better than one channel. The largest component of error arose from the limitation of six digits in the dividing circuit. The desired period differed from that obtainable by approximately 2 µs/cycle, or a steady slip of nearly two channels in a 40 min run. Omission of corrections for the Earth's axial rotation introduced a further slip of just less than one channel in the same time interval and in the same sense. If the value of $P_0 =$

Table 1. TIMING ERRORS OF THE OSCILLATORS

BBC carrier (200 kHz)	Maximum frequency error 5 parts in 10[10].
Local 5 MHz reference oscillator C	Manufacturer's specification, for any interval of 1,000 s 5 parts in 10[10]. Estimated error for a run of 5 h better than 5 parts in 10[9]. Measured accuracy of adjustment 1·4 parts in 10[9].
Running 1 MHz oscillator B	Maintained manually to better than 2 parts in 10[7].

1·33730113 s determined by Radhakrishnan et al. is correct, the slip will have been approximately fourteen channels in 40 min in the opposite sense.

An independent and direct determination of the cycling period of the analyser was made, using a pendulum clock. For 23 hours including a few hours during which the oscillator B was neither monitored nor adjusted, the value derived for the period in this way was found to be 1·337192 ± 0·000006 s. Most of the uncertainty in this determination arose from the comparison of the clock with BBC time signals to determine its rate.

We mention here that if the sweep frequency of −4·9 ± 0·5 MHz s[-1] observed at 80 MHz[1] is extrapolated to optical frequencies, the dispersion of arrival times for a delta-function optical pulse at the source, over the passband of the photometer (approximately 3500–6500 Å), would amount to only 4×10^{-13} s, which is utterly negligible in comparison with the resolution of the analyser of $2·6 \times 10^{-3}$ s.

If flashes of light were observable from CP 1919 a few adjacent channels in the Laben analyser would receive more counts than the average. Fig. 2 shows the result of a test in which light was reflected to the photometer by

Table 2. LIMIT OF DETECTION OF THE EQUIPMENT

	April 5, 1968		April 6, 1968	
Diaphragm dimensions, seconds of arc	10 × 10		30 × 30	
Run serial number	2	3	4	6
Time UT at start	02 17·5	03 18·5	02 06	03 15
at end	02 57·5	03 52	02 39	03 45
Total duration, min	40	33·5	33	30
Duration per channel, s	4·69	3·93	3·87	3·52
Sky count/s	810	750	5,100	4,800
Count/channel	3,800	2,950	19,750	15,150
r.m.s. noise in each channel	62	55	141	123
Sensitivity, counts/s on star, $B = 13·6$	2,000		3,000	
Limit of detection, magnitude at peak of pulses:				
assumption 1	18·05	17·95	17·35	17·45
assumption 2	18·3	18·2	17·6	17·7
assumption 3	18·3	18·2	17·6	17·7
Mean brightness:				
assumption 1	22·45	22·35	21·85	21·95
assumption 2	22·1	22·0	21·4	21·5
assumption 3	21·5	21·4	20·8	20·9

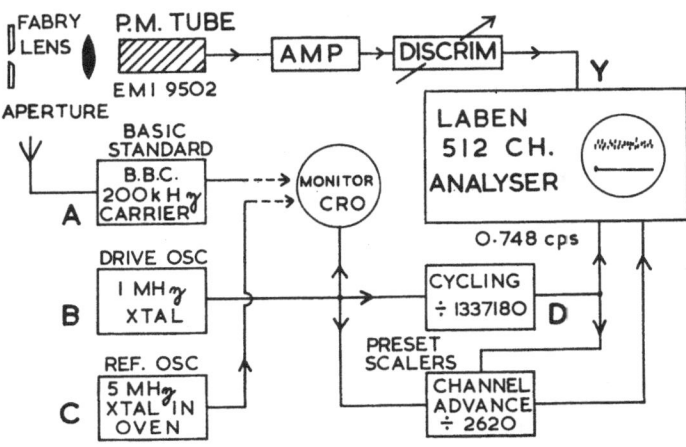

A. STANDARD. STAB 5 IN 10[10]

B. ADJUSTABLE, HELD TO "C" BY 1 IN 5×10[6]

C. LOCAL REF. SET TO 1 IN 7×10[8]
   SCALING ERROR AT D 1 IN 6·7×10[5]

EA. CH. 2·6 m.SEC
CAPACITY 65,000 COUNTS PER CHANNEL

Fig. 1. Block diagram of the equipment, showing the accuracy of the timing oscillators.

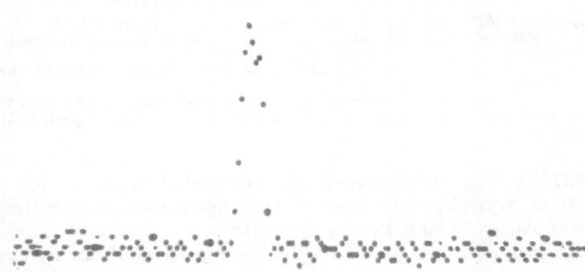

Fig. 2. The response of the analyser to a light flashing with a period of 1·337 s. Abscissae: channel number, equivalent to the phase of the variation. Ordinates: count accumulated, proportional to the brightness. The flashes lasted about 25 ms. In this figure the horizontal scale is enlarged and only about one-third of all the channels are displayed.

a mirror rotated with a period of 1·337 s. This figure was photographed from the cathode ray tube display of the analyser, with the horizontal (time) scale enlarged so that about one third of the total number of channels was displayed.

Four runs were made on CP 1919, during which no evidence for optical flashes was found. The details of the observations are given in Table 2, where the sensitivity of the equipment is computed for each run. A typical display is shown in Fig. 3, in which the time scale has been adjusted to include all 512 channels, and the vertical (number of counts) scale has been enlarged to show that

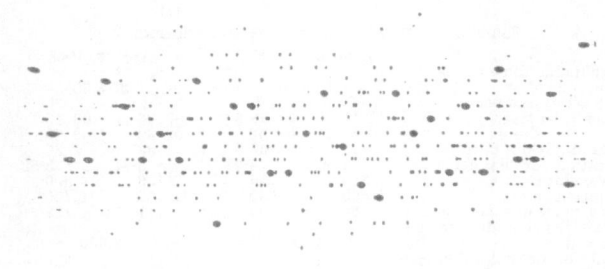

Fig. 3. Response of the analyser to CP 1919 (run 3). Abscissae and ordinates as in Fig. 2. In this figure all channels are displayed and the vertical scale is enlarged to show that no groups of adjacent channels have significantly more counts than the mean. In this display the counts in individual channels are rounded to the nearest multiple of 16, and every sixteenth channel is emphasized.

there is no sign of any systematic variation of brightness with phase.

The maximum brightness of flashes that we believe might go undetected depends on the assumed duration of the flashes and on the slip caused by inadequacies in the scalers or uncertainty in the period P_0. We assume that if the flashes last 16 ms (six channels) and the slip is two channels, so that the excess counts are distributed over eight channels, we should detect flashes if the eight channels have, on the average, two and a half times the r.m.s. variation of all counts above the mean. In Table 2 this is assumption 1. If the flashes last 35 ms and the slip of two channels causes the excess counts to be distributed over fifteen channels, we consider that an average excess of only twice the r.m.s. variation would be detectable. This is assumption 2. Finally, if the period $P_0 = 1·33730113$ s is correct, and the resultant slip of fourteen channels is added to a natural spread of 35 ms (thirteen channels), we again consider that an average excess of twice the r.m.s. variation would be detectable. This is assumption 3.

No optical flashes have been detected from CP 1919, and the mean brightness of flashes from any object within the error rectangle of Ryle and Bailey indicates that no more than 10 per cent of the light of the yellow object, assumed here to have magnitude $B = 19·0$, can be modulated in a manner similar to the radio signals, whichever of the two published heliocentric periods is correct. If the period of Hewish *et al.* is correct and the flashes are assumed to last 35 ms, not more than 6 per This evidence neither confirms nor refutes the identification of CP 1919 with the yellow object, but may help to elucidate some of its properties.

We are very grateful for the loan of the analyser and electronics from the Electronics and Applied Physics Division of AERE, Harwell, and we thank Dr P. Orman and Mr F. Pryce for designing the electronic system. We also thank Professor Redman for allocating time on the telescope at short notice.

Received May 3, 1968.

[1] Hewish, A., Bell, S. J., Pilkington, J. D. H., Scott, P. F., and Collins, R. A., *Nature*, **217**, 709 (1968), (Paper 1).
[2] Ryle, M., and Bailey, J. A., *Nature*, **217**, 907 (1968), (Paper 19).
[3] Hewish, A., *Meeting of the Royal Astronomical Society* (London, April 10, 1968).
[4] Johnson, H. L., *Astrophys. J.*, **116**, 640 (1952).
[5] Davies, J. G., Horton, P. W., Lyne, A. G., Rickett, B. J., and Smith, F. G., *Nature*, **217**, 910 (1968), (Paper 6).
[6] Radhakrishnan, V., Komesaroff, M. M., and Cooke, D. J., *Nature*, **218**, 229 (1968), (Paper 17).

26. Proper Motion Search for Pulsars ·

by

A. N. ARGUE
CECILIA M. KENWORTHY

The Observatories,
University of Cambridge

AN attempt has been made to identify optically three pulsars by proper motion surveys in the vicinity of the radio positions. An object at a distance of 100 pc having a velocity of 200 km s^{-1} perpendicular to the line of sight would have a proper motion of 0·4 arc s yr^{-1} and should therefore, if bright enough, be identifiable from a pair of photographs taken 10 or 20 yr apart.

The results have been completely negative. If we assume this is because pulsars are fainter than our limit of detection, then we can set upper limits to their lumino-

sity. Adopting the distances given by Lyne and Rickett[1] and the detection limits in column 4 of our table, we have derived the absolute magnitudes given in the final column. These are expressed on the B scale of the UBV system. CP 0950 has already been discussed by Bailey and Mackay[2], who derived $M_v = 16^m·5$. Our absolute magnitudes place CP 0834 and CP 1133 also in the extreme dwarf range or beyond, in conformity with current ideas[3].

For our first epoch photograph we reproduced portions

of the Palomar Sky Survey prints. The area was photographed on 35 mm film which was then contact printed on to a plate (Ilford N 50) suitable for clamping up in the measuring machine. The images had therefore passed through altogether five photographic processes before measurement: original negative in telescope, glass positive, paper print, film positive and finally glass negative. Distortions introduced at each stage will influence the final results, but these effects are not serious as can be seen from columns 5 and 6 of Table 1, which give the r.m.s. relative proper motion components obtained from the field stars. Considering that each proper motion is derived from only two measurements over a baseline of 13 to 17 yr, these r.m.s. values are satisfactorily low.

Our second epoch photographs were taken with the Cambridge 43/61 cm $f3\cdot8$ Schmidt telescope using blue sensitive (Ilford SRO) plates without filter. The limiting apparent magnitudes in column 4 were derived from

Table 1. LIMITS OF SURVEY, STANDARD DEVIATION IN PROPER MOTION COMPONENTS AND LUMINOSITIES

Field	Area (R.A. × Dec.)	No. of stars	Limiting apparent magnitude (B)	$<\mu_\alpha\cos\delta>$ (arc s yr^{-1})	$<\mu\delta>$	Absolute magnitude M_B
0834	$5\cdot6' \times 37'$	108	19	0·027	0·028	> 13·6
0950	$3\cdot4' \times 7\cdot1'$	6	17	0·021	0·013	—
1133	$7\cdot1' \times 38'$	49	19	0·028	0·040	> 15·6

estimates of image diameter on the Palomar prints using a graticule and the calibration published by Perek[4]. These limits may be in error by $\pm 0^m\cdot5$. They are the faintest we were able to achieve in a relatively bright evening sky under not very good conditions of transparency.

The areas searched had the dimensions given in column 2. These exceed the radio error rectangle dimensions by factors of about 2. The radio positions on which the areas were centred were taken from Bailey and Mackay[2] for CP 0950, and from Lyne and Rickett[1] for the other two. The numbers of stars measured are in column 3.

In not a single case did we find a significantly large relative proper motion. If pulsars are not more than 100 pc distant, they must either have remarkably low transverse velocities or their intrinsic luminosity in optical wavelengths (averaged in time) must be very small.

We thank Mr C. D. Mackay for the loan of film positives made from the Palomar Sky Survey prints.

Received June 7, 1968.

[1] Lyne, A. G., and Rickett, B. J., *Nature*, 218, 326 (1968), (Paper 10).
[2] Bailey, J. A., and Mackay, C. D., *Nature*, 218, 129 (1968), (Paper 20).
[3] Smith, F. G., *Nature*, 218, 720 (1968), (Paper 24).
[4] Perek, L., *Bull. Astr. Inst. Czechoslovakia*, 9, 39 (1958).

Theories

27. Pulsed Radio Sources

by
F. HOYLE
J. NARLIKAR
Institute of Theoretical Astronomy,
University of Cambridge

THE recent discovery[1] of rapidly pulsed radio sources and the further detailed observations[2] of the source CP 1919+22 raise remarkable problems. Undoubtedly the theory which requires the least change in our concepts is that of a pulsating white dwarf. As long ago as 1952, Mestel[3] pointed out that nuclear fuel in the outer regions of a white dwarf can cause an explosive instability. There is therefore no difficulty in understanding how an oscillatory state might be set up. It is also possible that oscillations could be self-sustaining.

The first problem this theory has to face is that the oscillation period of CP 1919+22 is remarkably short—1·337 s. Periods of more than 10 s were rather to be expected[4]. Recently, however, Thorne and Ipser[5] have argued that fundamental periods as low as 3 s might well occur in some white dwarfs, and they remark that an overtone period as low as 0·2 s is possible. This lower limit is close to the period of 0·253 s reported in this issue of Nature by Pilkington et al. for the source CP 0950+08. It seems therefore that the theory of white dwarfs is just about able to accommodate the shortest period so far observed. The accommodation is, however, uncomfortably tight, particularly because the time constant of

0·1 s used in the first Cambridge survey would have prevented the discovery of appreciably shorter periods.

The second problem is that the radio pulse occupies only a small fraction of the cycle, only about 40 ms in 1·33 s for CP 1919+22. Moreover, the pulse rise is exceedingly steep—the rise time is < 1 ms in this source. The disturbance leading to the radio pulse would have to be released at a sharply defined phase of each cycle and, to within a small margin, the same phase in every cycle. We could imagine a shock wave that reached the surface of the star always at a particularly defined phase, for example. The sharpness of the shock defines the steep rise of the pulse, while the pulse length is determined by the time difference between the arrival at the Earth of radiation from the centre of the disk and from the limb. Suppose the radiation from the centre of the disk arrives at time T. If R is the radius of the star, the radiation from the limb arrives at time $T + R/c$, thus giving a pulse length R/c. The pulse shape in the range $T \leq t \leq T + R/c$ can be determined in the following way. The radiation arriving at time t comes from the rim of a spherical cap of solid angle $2\pi(1 - \cos\theta)$ where θ is the angle between the radius vector to the rim and the direction to the

Earth. Clearly

$$t = T + (1 - \cos\theta)R/c \qquad (1)$$

If the radiation is emitted isotropically, the amount received in time dt is proportional to

$$2\pi \sin\theta\, d\theta . \cos\theta \qquad (2)$$

where the factor $\cos\theta$ arises from projection of the area normal to the line of sight. From equations (1) and (2) we see that the pulse shape is given by

$$\mathcal{J}(t) = K . \left(T + \frac{R}{c} - t\right) \qquad (3)$$

where K is a constant. Thus a triangular pulse shape with a steep rise and linear fall is expected for isotropic emission.

The pulse length of $CP\ 1919+22$ is about 40 ms and requires $R \sim 1\cdot2 \times 10^9$ cm. This is too large to correspond to the short period models of Meltzer and Thorne[4]. The difficulty is resolved if we argue that the observation really represents a series of pulses each of shorter length. The published pulse shapes[2] suggest that this might be so. In the shock wave model, this would imply a series of shock waves occurring over ~ 40 ms in each cycle of $CP\ 1919+22$.

The third problem is to understand how such a precise timing of the pulses can be combined with the great fluctuations in their amplitudes. How can such a very regular phenomenon be consistent with large amplitude variations, even over as short a time as a few minutes?

While these difficulties do not rule out the white dwarf explanation they make it reasonable to look for other explanations. One other possibility has recently been put forward by Saslaw, Faulkner and Strittmatter[6].

It is our chief concern in this communication to point out that a time constant of ~ 1 s follows from the collapse time of a supernova. According to previous ideas[8,9] a supernova is triggered by the endoergic nuclear reaction $^{56}Fe \rightarrow 13\alpha + 4n$ at the centre of an evolving massive star. The density and temperature at which collapse occurs have been calculated for various stellar masses. For masses in the range 10 $M_\odot$ to 30 $M_\odot$ the density was found to be $\sim 10^7$ g cm^{-3}. Because the implosion time is $[3\pi/32G\rho]^{1/2}$, this is $\sim 2/3$ s when we set $\rho = 10^7$ g cm^{-3}.

The fate of such a collapsing core has always been somewhat uncertain. Collapse into a gravitational singularity is one possibility, but it has always been thought more probable that the core will contrive to "bounce", in which case it will oscillate back to more or less the radius it had at the onset of collapse, about 10^9 cm. The time for a complete oscillation is essentially twice that estimated in the previous paragraph, namely, $\sim 4/3$ s, very close to the observed period of $CP\ 1919+22$, and close to $1\cdot27$ s for $CP\ 0834+07$ and to $1\cdot19$ s for $CP\ 1133+17$, but longer than the period $0\cdot253$ s of $CP\ 0950+08$. A shorter observed period, however, does not necessarily present a serious difficulty, because damping processes—for example, neutrino-emission—shorten the theoretical period.

At this stage we can follow one or other of two lines of development.

(a) The object continues to oscillate between a maximum radius $R_{max} \sim 10^9$ cm and a minimum radius $R_{min} \ll R_{max}$. The value of R_{max} determines the pulse repetition rate, while the phase of R_{min}—corresponding to the greatest compression of the object—determines the moment of pulse emission in relation to the cycle. The pulse could arise from the compression of a magnetic field, for example[7].

(b) The explosion of nuclear fuel in the outer regions of the star[8] reduces the total mass to less than the upper limit of a stable white dwarf. After the supernova explosion has taken place, the core then springs back to a white dwarf configuration.

Possibility (a) is new, while (b) connects with the white dwarf theory. It is not necessarily an objection to (a) that oscillations with $R_{min} < R_{max}$ would rapidly be damped—for example, by neutrino emission—because it is possible for the object to divide into a number of pieces which maintain constant high density, the oscillation being confined to the pieces—the pieces moving inwards together and outward together. Indeed, the theory of Saslaw, Faulkner and Strittmatter[6] can be incorporated into this possibility because it can be regarded as a case in which the core possesses appreciable angular momentum. In such a case it is possible for division to occur into two dense pieces in binary motion about each other.

The advantage of (b) in relation to the white dwarf, theory is that it provides a natural explanation of the periodicity of ~ 1 s. As we have already noted, periods of this order represent the shortest periods which it is possible for white dwarfs to have. Such stars are at the margin of stability. Increase the mass slightly and collapse occurs into the kind of object considered in (a). By working from above, however, by reducing the mass towards stability, we obtain an understanding of why the pulsed sources tend to lie near the margin of stability.

The dynamical energy in case (a) is $\sim 10^{54}$ ergs (ref. 7), enormous compared with the energy requirements of the pulsed sources, which amount to only $\sim 10^{28}$ ergs s^{-1}, at any rate so far as the radio emission is concerned. This would seem to us a disadvantage of (a) because we would expect such a large energy source would manifest itself in a much more dramatic form. On the other hand, it is perhaps easier to understand the rapid variations of shape and amplitude of the pulses in case (a). This is especially so if the object divides into a number of pieces. However well the pieces maintain an organized in-and-out motion, we cannot expect the detailed arrangement of the pieces to be exactly the same at minimum radius, $R_{min} \ll R_{max}$, in each cycle. Variations over a few cycles or even from cycle to cycle might be expected and could provide the explanation of the remarkable changes of amplitude of the radio pulses.

Next we remark on the likely composition of the white dwarf corresponding to case (b). Hoyle and Fowler[8] considered that most of the helium produced by $^{56}Fe \rightarrow 13\alpha + 4n$ did not recombine. On this view, the white dwarf would be largely helium, as suggested by Thorne and Ipser[5]. Some helium would combine to form ^{12}C and higher elements, however, while the free neutrons would eventually decay to protons. It is interesting that $^{12}C(p,\gamma)^{13}N(\beta^+)^{13}C$ provides a source of ^{13}C. Together with the α particles, ^{13}C has the very temperature sensitive reaction, $^{13}C(\alpha,n)^{16}C$. This reaction has been considered as a source of instability in stars. The possibility suggests itself that this same reaction might be the source of white dwarf oscillations.

Finally, we note that an active lifetime of at least 3×10^6 years is necessary for the supernova residue in both cases (a) and (b). In 3×10^6 years we expect about 10^5 supernovae to occur of which ~ 10 would lie within 100 pc. This is of the order of the density of pulsed sources suggested by the Cambridge survey. Because there are $\sim 10^{14}$ cycles in 3×10^6 years, it is therefore necessary that the secular change of period per period should not be more than about one part in 10^{14}. This indeed seems to be the case (F. G. Smith, private communication).

Received April 8, 1968.

[1] Hewish, A., Bell, S. J., Pilkington, J. D. H., Scott, P. F., and Collins, R. A., Nature, 217, 709 (1968), (Paper 1).

[2] Davies, J. G., Horton, P. W., Lyne, A. G., Rickett, B. J., and Smith, F. G., Nature, 217, 910 (1968), (Paper 6).

[3] Mestel, L., Mon. Not. Roy. Astro. Soc., 112, 598 (1952).

[4] Meltzer, D. W., and Thorne, K. S., Ap. J., 145, 514 (1966).

[5] Thorne, K. S., and Ipser, J. R., Ap. J. Letters (in the press).

[6] Saslaw, W. C., Faulkner, J., and Strittmatter, P. A., Nature, 217, 1222 (1968), (Paper 42).

[7] Hoyle, F., Narlikar, J. V., and Wheeler, J. A., Nature, 203, 914 (1964).

[8] Hoyle, F., and Fowler, W. A., Ap. J., 132, 565 (1960).

[9] Fowler, W. A., and Hoyle, F., Ap. J. Suppl., 91 (1964).

28. Stability and Radial Vibration Periods of the Hamada–Salpeter White Dwarf Models

by
JOHN FAULKNER
JOHN R. GRIBBIN
Institute of Theoretical Astronomy,
University of Cambridge

THE recent discovery by Hewish et al.[1,2] of rapidly pulsing radio sources with periods of $\sim 1\cdot337$, $1\cdot274$, $1\cdot188$ and $0\cdot2531$ s has already provoked a considerable amount of activity among theoretical astrophysicists. Hewish et al. sought an explanation in terms of the radial vibrations of white dwarf or neutron star models as calculated by Meltzer and Thorne[3] (hereafter MT). The apparent failure of the MT white dwarfs to pulsate in the fundamental mode with a period less than ~ 8 s led Hewish et al. to suggest tentatively that neutron stars might possibly be responsible. While several authors have proposed other, more or less bizarre theories[4-6], we shall show in this article that the white dwarf explanation may have been dismissed prematurely. We have found that the fundamental periods for the reasonably realistic white-dwarf models of Hamada and Salpeter[7] can be as low as $\sim 1\cdot79$ to $\sim 2\cdot05$ s, the actual value being somewhat dependent on the composition. Furthermore, we shall show that similar results would have been obtained by Meltzer and Thorne but for an error in their specification of γ, the generalized ratio of specific heats, just where this is most crucial. Finally, the relevance of our results to the CP sources themselves will be briefly discussed.

The Hamada–Salpeter white dwarfs[7] were originally constructed using the Salpeter zero temperature equation of state[8] and Newtonian stellar structure equations. The stability properties are, however, intimately bound up with general relativistic effects in the region of particular interest (central density in the range $\sim 10^{10}\text{--}10^{11}$ g cm^{-3}). We have therefore chosen to construct our models using the relativistic procedure throughout. The standard relativistic structure equations and the pulsation equation have been given by Chandrasekhar[9] and Bardeen, Thorne and Meltzer[10], and we shall as far as possible use the same notation as these authors.

Hamada and Salpeter considered that white dwarfs might be formed with a variety of compositions and that non-uniformities could arise at low temperatures through the operation of inverse β decays and pycnonuclear reactions. As well as building models of uniform composition, we have also built non-uniform models using the Hamada–Salpeter prescription. Thus our non-uniform ^{4}He$_2$ models contain ^{12}C$_6$ where $\rho \gtrsim 8 \times 10^8$ g cm^{-3}, and ^{24}Ne$_{10}$ where $\rho \gtrsim 6 \times 10^9$ g cm^{-3}, while our "equilibrium" models contain ^{56}Fe$_{26}$ for log $\rho < 7\cdot15$, and heavier, more neutron-rich nuclei[8] for higher densities here considered ($11\cdot28 < \log \rho \leqslant 11\cdot40$). For purposes of comparison we have also computed models having pure ^{4}He$_2$ and ^{120}Sr$_{38}$ throughout, although neither of these compositions can be realistic at high or low densities respectively. Finally, as the Salpeter equation of state depends on both the charge Z of the nuclei and the mean atomic weight per electron, μ_e, we computed models for a fictitious composition (corresponding to ^{24}O$_8$) as a means of examining the Z dependence by comparison with the ^{120}Sr$_{38}$ results.

Working with the usual form of the interior Schwarzschild metric[9,10], we find that M (the total mass-energy)

and R (the surface radial co-ordinate) are in close agreement with the corresponding mass and radius of the Hamada–Salpeter models. This is to be expected, as an examination of the solutions shows that e^ν and e^λ differ from unity by at most a few per cent in the models of highest central density here considered.

In the pulsation calculations we have assumed, following MT, that there are two main time scales involved in the problem: (i) t_{relax}, the relaxation time-scale for nuclear reactions, and (ii) τ, the free oscillation period of the star. For $\tau \gg t_{\text{relax}}$, the nuclear species always remain at the local equilibrium values, necessarily changing during the long time-scale oscillation. This leads to instability over a very wide range of central density, as Meltzer and Thorne have shown. Again following Meltzer and Thorne, we have assumed that $t_{\text{relax}} \gtrsim 10^{10}$ yr, however. Even the present astronomically unprecedented accuracy of the CP periods cannot rule out a drift rate with this timescale, and we shall examine oscillations with periods $\tau \ll t_{\text{relax}}$ under the assumption that such metastable states are, for present purposes, stable.

When $\tau \ll t_{\text{relax}}$, we therefore assume that the nuclear abundances remain fixed throughout the oscillations and that γ is given in the MT notation by

$$\gamma = \frac{\rho^* + p^*}{p^*} \left. \frac{\partial p^*}{\partial \rho^*}\right|_{\text{fixed, local nuclear abundances}} \tag{1}$$

The fundamental pulsation periods of our models are given in Table 1, together with the first overtone periods for the more realistic compositions (we do not list higher overtones, as these are given, to within a few per cent, by $\tau_n \propto 1/(n+1)$ for the nth overtone, $n \geqslant 1$). The periods of some fundamentals and of the equilibrium composition first overtone are shown in Fig. 1, which may be compared directly with Fig. 4 of Meltzer and Thorne.

The first and most obvious difference between our results and the MT results is that, for any composition, our models remain stable until log $\rho_c \gtrsim 10\cdot35$, that is, $\rho_c \gtrsim 2\cdot3 \times 10^{10}$ g cm^{-3}. Our periods are also considerably lower, and, over a fairly wide range of central densities, differ from the three longer CP periods by only moderate factors $\lesssim 2$. Note that the shortest period, closest to the CP observations, is only obtained with the unrealistic pure ^{120}Sr$_{38}$ composition, and that the more realistic equilibrium abundance gives $\tau_{\min} = 1\cdot79$ s. It would, however, be most intriguing if the observations were hinting that high Z, neutron-rich material is present in the centres of highly evolved white dwarfs. We hope that this result, meagre as it is, may stimulate others to study reactions at these high densities. More studies along the lines of Wolf[11] are clearly indicated.

The following question obviously arises: how can our results differ so much from those of Meltzer and Thorne ? Our equilibrium models should be fairly close in properties to the MT models at these densities. The answer appears to lie in the MT specification of γ.

After we had discovered this, Professor W. A. Fowler

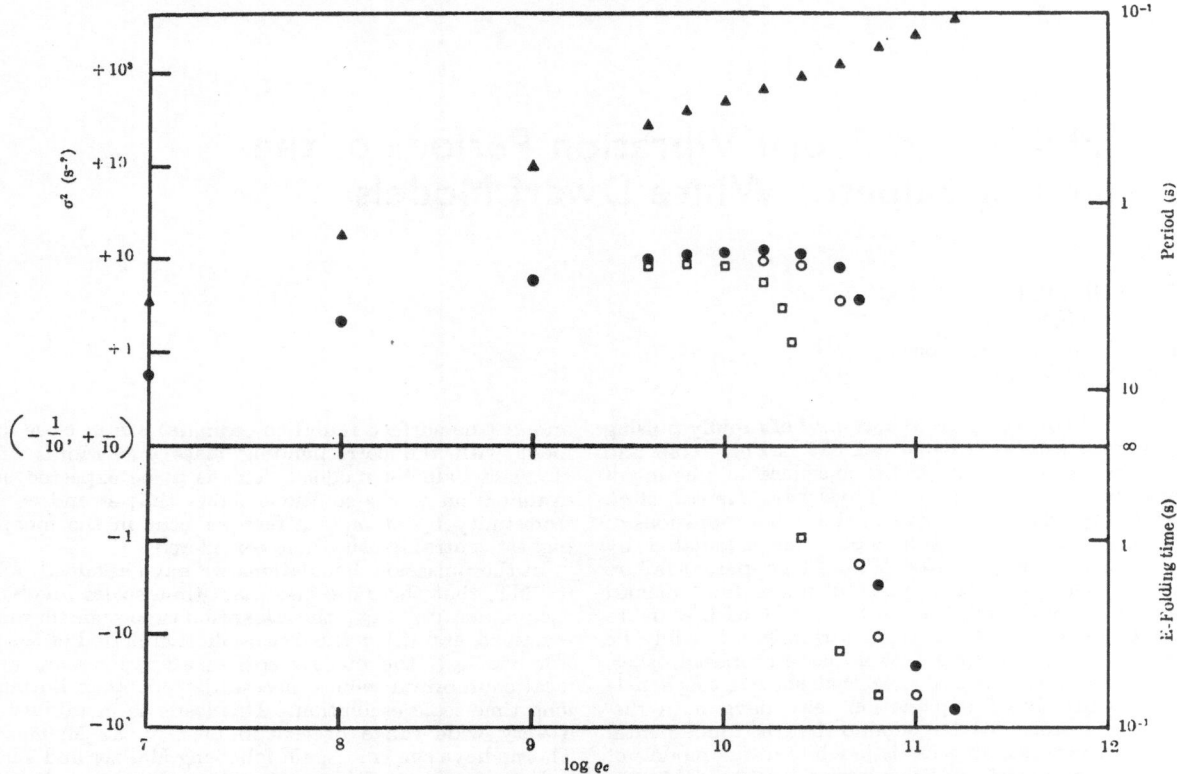

Fig. 1. Values of σ^2, period and e-folding time for three fundamentals and one first overtone. □, Pure $^4He_2$, fundamental; ○, inhomogeneous $^4He_2$, fundamental; ●, equilibrium composition, fundamental; ▲, equilibrium composition, first overtone.

kindly pointed out that the question of stability for a pure helium, purely Chandrasekhar equation of state, white dwarf model had been settled some time ago by Chandrasekhar and Tooper[12], who found that instability was reached when $\rho_c = 2 \cdot 328 \times 10^{10}$ g cm^{-3}. While the matter is thus not in doubt, we offer the following argument as indicating rather well the region where instability may be expected.

When $\tau \ll t_{\text{relax}}$ Meltzer and Thorne assumed that in the density range $10^7 \lesssim \rho \lesssim 10^{12}$, where relativistically degenerate electrons provide almost all the pressure, γ could be put equal to 4/3. Unfortunately, the Newtonian stability condition is $\gamma - 4/3 \geqslant 0$ (or some suitable average, $\overline{\gamma - 4/3} \geqslant 0$ when γ is variable), whereas in the relativistic case a post-Newtonian expansion shows that the relevant criterion is

$$\gamma - 4/3 \geqslant 0 \left(\frac{GM}{Rc^2} \right)$$

Table 1. PERIODS OF RADIAL OSCILLATIONS (s)

Log ρ_c	Fundamental mode Homogeneous models			Inhomogeneous models		First overtone	
	$^4He_2$	$^{16}O_8$	$^{120}Sr_{38}$	"$^4He_2$"	"Equilib."	"$^4He_2$"	"Equilib."
7·0	8·60			8·60	8·22	3·44	3·30
8·0	4·49			4·49	4·31	1·46	1·48
9·0	2·71	2·36	2·27	2·71	2·56	0·620	0·633
9·6	2·19	1·82	1·74	2·19	2·01	0·360	0·383
9·8	2·14	1·70	1·62	2·13	1·90	0·322	0·325
10·0	2·20	1·61	1·53	2·09	1·83	0·291	0·283
10·2	2·65	1·58	1·48	2·05	1·79	0·266	0·245
10·4	*1·05	1·65	1·49	2·16	1·84	0·234	0·213
10·6	*0·257	2·07	1·69	3·32	2·22	0·198	0·180
10·8	*0·151	*0·504	3·62	*0·311	*0·582	0·162	0·149
11·0	*0·100	*0·176	*0·218	*0·151	*0·217	0·130	0·127
11·2	*0·070	*0·107	*0·119	*0·094	*0·126	0·102	0·106
11·4	*0·050		*0·077	*0·062	*0·096	0·080	0·094

* Denotes an unstable model, for which the e-folding time is given.

The curve of inhomogeneous period as a function of ρ_c is slightly "wavy". The waves are correlated with the passage of a composition discontinuity over the region where e$^\lambda$ (or GM/rc^2) achieves a maximum in the model, and are somewhat dependent on the rapidity of the change-over. The compositions here have been smoothed over a range of $\sim 0 \cdot 12$ in log $_{10}P$.

The integrations were performed with successively reduced step-lengths, as a check on accuracy. We estimate these figures to be accurate to about 1 or 2 per cent.

The location of the instability may then be understood as follows.

At high densities the Salpeter pressure reduces to a constant multiple of the Chandrasekhar value. The latter is conveniently expressed in terms of the degeneracy parameter x, where $\rho \sim 10^6 \mu_e x^3$. The correct expansion of the Chandrasekhar function $f(x)$[13] for the degenerate pressure at high densities gives

$$P \propto f(x) = 2x^4 - 2x^2 + 0 (\ln_e x) \qquad (2)$$

It may readily be shown that this leads to

$$\gamma = \frac{4}{3} + \frac{2}{3} x^{-2} + 0 (x^{-4} \ln_e x) \qquad (3)$$

the relativistic corrections to the γ of equation (1) being negligible.

We thus expect stability provided that, in the mean,

$$\gamma - \frac{4}{3} \sim \frac{2}{3} \cdot 10^4 \left(\frac{\mu_e}{\rho} \right)^{2/3} \gtrsim \frac{GM}{Rc^2} \qquad (4)$$

The function M/R reaches a maximum inside the star, while $\rho^{-2/3}$ is monotonic. A good indication of the critical region is therefore given by setting the two sides of the inequality (4) equal at the point where M/R reaches its maximum, that is, at the point where $M = 4\pi R^3 \rho$, or, alternatively,

$$\rho (R) = \frac{1}{3} \overline{\rho} (R) \qquad (5)$$

Let

$$\rho (R) = \lambda \rho_c \qquad (6)$$

at this point. Then $R = (M/4\pi\lambda \rho_c)^{1/3}$ and we obtain, after substituting in equation (4) with $\mu_e = 2$,

$$\lambda \rho_c = \frac{10^{32}}{\pi^{1/3} M^{2/3}} \qquad (7)$$

In the present range of central densities, $\rho_c/\overline{\rho}$ (surface) ~ 30, which places the models between polytropes of index 3 and 2·5, as one might expect. A glance at the polytropic tables shows that, for $\rho/\bar{\rho} \sim 1/3$, $\lambda \sim 1/6$ and $M \sim 2/3\,M_{total}$ represent good compromises. As we are close to the limiting mass, this means $M \sim M_\odot \sim 2 \times 10^{33}$ g. Thus

$$\rho_c \sim \frac{6 \times 10^{10}}{(4\pi)^{1/3}} \simeq 2\cdot6 \times 10^{10} \qquad (8)$$

This result is, of course, no better than the rough arguments which have gone into it. We are grateful to Professor Fowler for showing us that this estimate can be put on much firmer ground by the use of his HDBE method (see, for example, ref. 14), which with suitable changes reproduces fairly well the position at which instability occurs and the values of the periods which we obtain. As a final check on the MT results, computations were performed with the Hamada–Salpeter equilibrium models in which, during the pulsation, γ was artificially forced equal to $4/3$ for densities exceeding 10^7 g cm^{-3}. The periods or e-folding times were then within a few per cent of the MT values, with instability occurring, as in their models for ρ_c slightly in excess of 10^9 g cm^{-3}, where the destabilizing effect of the $\gamma = 4/3$ region overcomes the stabilizing effect of the matter with $\rho < 10^7$ g cm^{-3}. We conclude that the MT results are therefore erroneous for $\rho_c \gtrsim 10^7$ g cm^{-3}.

Finally, what is the relevance of this to the CP sources ? The close approach of the minimum periods in Table 1 to the three larger CP periods makes it attractive to suppose that these objects are in fact white dwarfs vibrating in their fundamental radial mode. It might also be thought that the most probable periods to be observed would be those corresponding to the flat portions of the curve in Fig. 1. This would not in fact be the case for the hypothetical "pure" compositions. In accordance with the $M(R)$ theorem[10] we find that M reaches a maximum where instability occurs. Where the period has its minimum (in fact where $\rho_c(\tau_{min}) \sim 1/4\,\rho_c(\tau = \infty)$, in agreement with Fowler's HDBE method) the curve of $M(\rho_c)$ is still very flat, giving small mass ranges (~ 2 per cent) for probable observation. The pycnonucleated models fare much better, the $M(R)$ theorem then being inapplicable. In fact, τ lies within 10 per cent of τ_{min} for about a 20 per cent range of mass. The latter models do, however, have a formation problem, as the mass is decreasing with increasing ρ_c in this range, and we have to decide whether a given mass will form a white dwarf

at the higher or lower of its allowed central densities. Curiously enough, this seems to call for a forced, implosive collapse.

On the other hand, it might be thought rather surprising if surface nuclear reactions were to excite the fundamental mode in preference to some high overtone. (The argument is less persuasive if low temperature, high density reactions are responsible, as our work may indicate.) In our range in which τ lies within 10 per cent of τ_{min}, our overtones have periods in the range $\sim 0\cdot20$ to $0\cdot35$ s, the correct range for the most rapidly pulsing CP source. One is attracted to the idea that a given excitation mechanism operating in similar objects might ultimately result in, at worst, adjacent overtones, and the fact that the nth harmonic period is proportional to $1/(n+1)$ for $n \geqslant 1$ mitigates strongly against high overtones. In fact it would be an odd coincidence, to say the least, were nature to contrive to present us with such transparently associated periods. Clearly more extensive investigations of excitation mechanisms, finite temperature effects, rotationally aided stability and so on, are called for.

We thank Professors R. F. Christy, W. A. Fowler and F. Hoyle and Dr P. A. Strittmatter for many helpful discussions, and J. Skilling for communicating his results on the fundamental modes before publication. One of us (J. R. G.) is indebted to the Science Research Council and University College, Cambridge, for research studentships.

Received May 15, 1968.

[1] Hewish, A., Bell, S. J., Pilkington, J. D. H., Scott, P. F., and Collins, R. A., *Nature*, **217**, 709 (1968), (Paper 1).
[2] Pilkington. J. D. H., Hewish, A., Bell, S. J., and Cole, T. W., *Nature*, **218**, 126 (1968), (Paper 2).
[3] Meltzer, D. W., and Thorne, K. S., *Ap. J.*, **145**, 514 (1966).
[4] Saslaw, W. C., Faulkner, J., and Strittmatter, P. A., *Nature*, **217**, 1222 (1968), (Paper 42).
[5] Ostriker, J. P., *Nature*, **217**, 1227 (1968), (Paper 39).
[6] Burbidge, G. R., and Strittmatter, P. A., *Nature*, **218**, 433 (1968), (Paper 43).
[7] Hamada, T., and Salpeter, E. E., *Ap. J.*, **134**, 683 (1961).
[8] Salpeter, E. E., *Ap. J.*, **134**, 669 (1961).
[9] Chandrasekhar, S., *Ap. J.*, **140**, 417 (1964).
[10] Bardeen, J. M., Thorne, K. S., and Meltzer, D. W., *Ap. J.*, **145**, 505 (1966).
[11] Wolf, R. A., *Phys. Rev.*, **137**, B1634 (1965).
[12] Chandrasekhar, S., and Tooper, R. F., *Ap. J.*, **139**, 1396 (1964).
[13] Chandrasekhar, S., *An Introduction to the Study of Stellar Structure*, second ed., 360 (Dover, 1957).
[14] Fowler, W. A., *Ap. J.*, **144**, 180 (1966).

29. Pulsation Periods of Rotating White Dwarfs

by

B. R. DURNEY
High Altitude Observatory,
Boulder, Colorado

J. FAULKNER
J. R. GRIBBIN
Institute of Theoretical Astronomy,
University of Cambridge

I. W. ROXBURGH
Department of Mathematics,
Queen Mary College, University of London

THE discovery by Hewish *et al.*[1,2] of pulsating radio sources with periods in the range of $0\cdot25$ to $1\cdot34$ s has stimulated renewed interest in the oscillation properties of highly condensed stars in the search for objects with fundamental periods of this order. Such calculations were made by Roxburgh and Durney[3] in 1966 as part of an

investigation into the stability of rotating white dwarfs and we recall their results in this paper. Their investigation was primarily concerned with determining at what point white dwarfs became dynamically unstable, and in particular in determining the effect of uniform rotation on the onset of instability. The calculations were done

Table 1. EVALUATION OF PULSATION PERIODS FOR UNIFORM ROTATION

Log ϱ	y_0	$\varrho_c/\bar{\varrho}$	Σ_0^2	Σ_1^2	β_{max}	$\Sigma_{\beta^2 max}$	(Period)$_R$	(Period)$_{non\ rot.}$
9·0	8·062	23·2	3·290	$0·152 \times 10^3$	$9·23 \times 10^{-3}$	4·89	2·24	2·73
9·45	11·36	28·0	4·638	0·453	7·58	8·07	1·75	2·30
9·9	16·03	32·5	5·219	1·334	6·53	13·92	1·50	2·17
10·35	22·65	37·0	0·562	3·922	5·74	22·80	1·04	6·62
10·80	32·02	41·5	− 27·30	11·31	5·11	30·49	0·90	—
11·25	45·27	45·0	− 144·9	32·5	4·72	8·50	1·70	—
11·70	64·01	49·0	− 529·7	94·9	4·33	− 118·8		

Note: The table uses information from Tables 1 and 2 of ref. 3; y_0 is the central value of the degeneracy parameter used in that paper.

for pure helium white dwarfs with a Chandrasekhar equation of state and included the effects of general relativity and uniform rotation. Calculations with more realistic equations of state, but without the effect of rotation, have recently been undertaken by Meltzer and Thorne[4] and corrected by Faulkner and Gribbin[5], but these authors have not found periods as short as those observed. We remark that the calculations by Faulkner and Gribbin show that simple degenerate models of pure helium have longer periods than models with more sophisticated compositions and equations of state, by a factor of up to 3/2, so that if we find sufficiently short periods for simple helium rotating white dwarfs, we expect the situation to be even better for more realistic models.

The equilibrium equations and the equations governing small pseudo-radial oscillations about equilibrium are given by Roxburgh and Durney[3], who give the oscillation frequency σ in terms of the dimensionless frequency Σ, defined by

$$\Sigma^2 = \frac{\sigma^2 q c^2}{32 \pi A G}$$

where $q = M_\odot/\mu H$, $A = \pi M_e^4 c^5/3h^3$, and G and c are the constant of gravitation and the velocity of light. The results for Σ^2 as a function of ϱ_c, the central density, are expressed in the form

$$\Sigma^2 = \Sigma_0^2 + \beta \Sigma_1^2$$

where Σ_0 is the frequency in the absence of rotation and $\beta \Sigma_1^2$ is the change due to a uniform rotation Ω, and

$$\beta = \frac{\Omega^2}{2\pi G \varrho_c}$$

The values of Σ_0^2 and Σ_1^2 are given in Table 1 for a range of values of ϱ_c.

To estimate the maximum change in frequency due to rotation, we need to calculate the maximum permitted value of β. This is given by the condition that centrifugal force balances gravity at the equator of the star, and depends on the central condensation of the star; a suitable measure of which is $\varrho_c/\bar{\varrho}$, where ϱ_c is the central density and $\bar{\varrho}$ is the mean density. This is given in Table 1. A simple calculation demonstrates the dependence of β on $\varrho_c/\bar{\varrho}$. If R_e is the equatorial radius of the star of mass M, then for maximum rotation

$$\Omega^2 R_e \approx \frac{GM}{R_e^2}$$

treating the mass distribution as spherically symmetric. If R_0 is the mean radius of the star, so that $M = 4\pi\bar{\varrho}R_0^3/3$, then

$$\frac{\Omega^2}{2\pi G \bar{\varrho}} = \frac{2}{3}\left[\frac{R_0}{R_e}\right]^3$$

For a centrally condensed star (R_0/R_e) is independent of the central condensation, the surface being just the simple Roche surface, so that

$$\beta = \frac{\Omega^2}{2\pi G \varrho_c} = \frac{2}{3}\left[\frac{R_0}{R_e}\right]^3 \left[\frac{\bar{\varrho}}{\varrho_c}\right] \propto \left[\frac{\bar{\varrho}}{\varrho_c}\right]$$

Calculations by Monaghan and Roxburgh[6] give the value of β for the case $(\varrho_c/\bar{\varrho}) = 54$ (a polytrope of index 3), as $3·95 \times 10^{-3}$ so that the maximum values of β for other degrees of central condensation are readily calculated. The results of such calculations are given in Table 1. With these values of β the corresponding maximum value of Σ_R^2 for the rotating case can be determined and hence

the corresponding periods. These are also given in Table 1, where we see that periods as short as 0·90 s can be obtained. This is less than the period of three of the four pulsed radio sources.

Non-uniform rotation with Ω higher in the central regions than the equatorial regions permits a higher effective value of β and hence an even smaller period. A very crude estimate of this effect can be made as follows. The maximum rotation is given by considerations of dynamical stability and unpublished investigations by Roxburgh show that the ratio of kinetic to gravitational energy is at most 2/15. We shall define the mean angular velocity $\bar{\Omega}$ by writing the kinetic energy in rotation as

$$T = \frac{1}{2} I \bar{\Omega}^2 = \frac{2}{15} \times \left[\frac{3}{5-n} \frac{GM^2}{R}\right]$$

where the last term in brackets is the gravitational energy of a polytrope of index n. Now the moment of inertia, I, can be expressed as

$$I = KMR^2$$

where K and n are known for a given $\varrho_c/\bar{\varrho}$ from detailed integrations. A little rearranging now gives, for the models considered here,

$$\bar{\beta}_{max} = \frac{\bar{\Omega}^2_{max}}{2\pi G \varrho_c} \approx \frac{0·2}{K}\left[\frac{\bar{\varrho}}{\varrho_c}\right]$$

The values of K, $\bar{\beta}_{max}$, Σ^2_{max} and the period are given in Table 2. Extremely short periods of less than 0·1 s are obtained. Although these estimates for non-uniform rotation cannot be considered as exact, they do suggest that there should be no difficulty in obtaining very short periods from white dwarfs with a large amount of rotational energy.

Table 2. EVALUATION OF PULSATION PERIODS FOR NON-UNIFORM ROTATION

Log ϱ_c	$\varrho_c/\bar{\varrho}$	K	β_{max}	Σ^2_{max}	Period (s)
9·0	23	0·110	$8·39 \times 10^{-2}$	16·1	1·24
9·45	28	0·102	$7·43 \times 10^{-2}$	38·2	0·81
9·9	32·5	0·096	$6·70 \times 10^{-2}$	94·5	0·51
10·35	37	0·091	$6·31 \times 10^{-2}$	247	0·32
10·80	41·5	0·087	$5·87 \times 10^{-2}$	637	0·20
11·25	45	0·083	$5·69 \times 10^{-2}$	1,700	0·12
11·70	49	0·080	$5·41 \times 10^{-2}$	4,600	0·07

We also remark that the maximum value of 2/15 for the ratio of rotational and gravitational energy is just what one might expect from condensing a main sequence star of roughly the limiting mass. If white dwarfs with mean densities of order 10^9 g cm^{-3} form from stars with mean densities of order unity, the corresponding initial value of the energy ratio would only be $\sim 10^{-4}$, corresponding to equatorial velocities of order 30 km s^{-1}. From Allen[7], main sequence stars of $\sim 1·4$ M$_\odot$ are of spectral type F5 and have mean rotational velocities $\bar{V}_e \sim 25$ km s^{-1}. While not conclusive, the agreement is nevertheless most striking.

Received June 21, 1968.

[1] Hewish, A., Bell, S. J., Pilkington, J. D. H., Scott, P. F., and Collins, R. A., *Nature*, 217, 709 (1968), (Paper 1).

[2] Pilkington, J. D. H., Hewish, A., Bell, S. J., and Cole, T. W., *Nature*, 218, 126 (1968), (Paper 2).

[3] Roxburgh, I. W., and Durney, B. R., *Z. Astrophys.*, 64, 504 (1966).

[4] Meltzer, D. W., and Thorne, K. S., *Ap. J.*, 145, 514 (1966).

[5] Faulkner, J., and Gribbin, J. R., *Nature*, 218, 734 (1968), (Paper 28).

[6] Monaghan, J. J., and Roxburgh, I. W., *Mon. Not. Roy. Astro. Soc.*, 131, 13 (1965).

[7] Allen, C. W., *Astrophysical Quantities*, second ed. (University of London, Athlone Press, 1963).

30. Radial Oscillation Periods for Hamada–Salpeter Models of White Dwarfs

by

J. SKILLING

Mullard Radio Astronomy Observatory,
Cavendish Laboratory,
University of Cambridge

RADIAL pulsations of white dwarfs may be responsible for the radio pulses discovered by Hewish *et al.*[1], and so quantitative computations of the oscillation periods have become important to observational astronomers. The calculations of Meltzer and Thorne[2] are based on the Harrison–Wheeler–Wakano equation of state, which describes cold matter continuously catalysed to the composition of lowest free energy. As these authors point out, nuclear thermodynamic equilibrium is unlikely to be

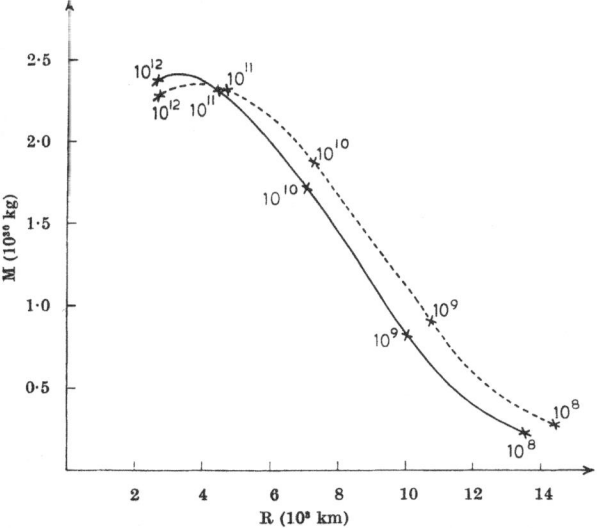

Fig. 1. The mass-radius curve for the Harrison-Wheeler-Wakano equation of state. ———, Newtonian results obtained by the author; - - -, general relativistic results. The parameters on the curves are central density in kg m⁻³.

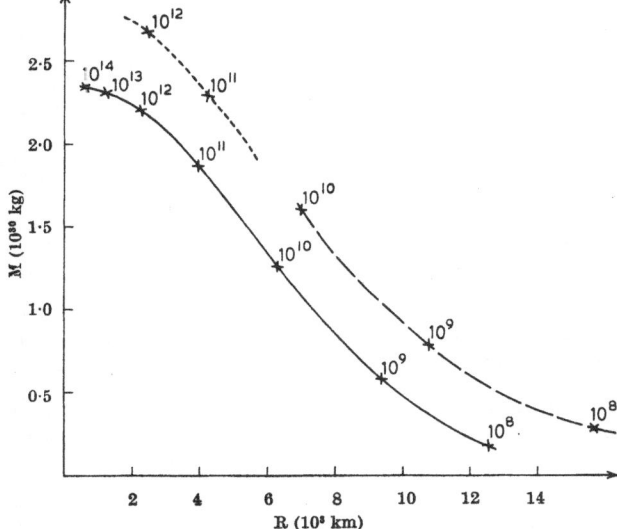

Fig. 2. The mass-radius curve for the Hamada-Salpeter equation of state. ———, Iron; - - -, carbon; — — —, helium. The parameters on the curves are central density in kg m⁻³.

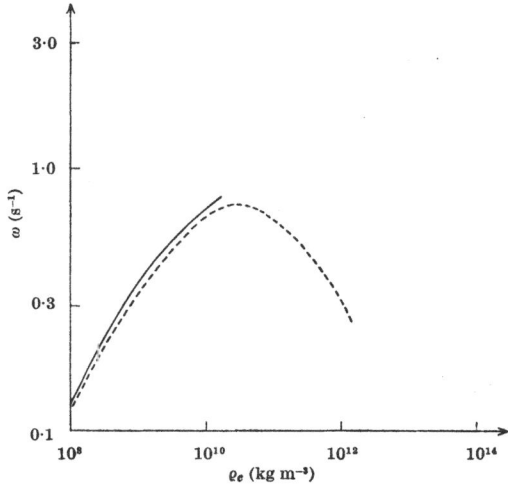

Fig. 3. Oscillations of Harrison-Wheeler-Wakano models. ———, Newtonian results obtained by the author; . . ., general relativistic results obtained by Meltzer and Thorne[2].

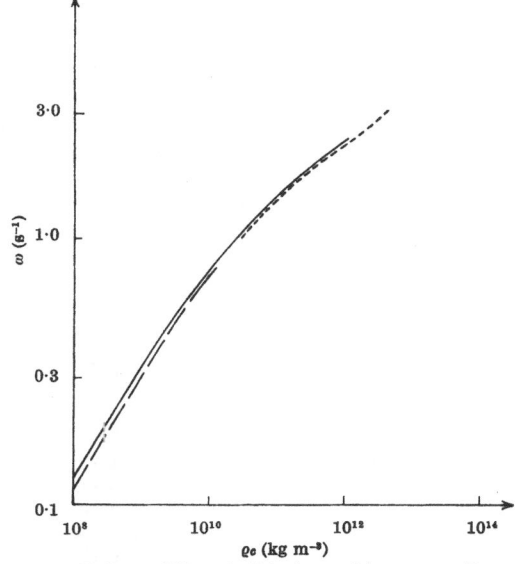

Fig. 4. Oscillations of Hamada-Salpeter models. ———, Iron; - - - carbon; — — —, helium.

reached in real white dwarfs; real white dwarfs are probably better described by the Hamada–Salpeter[3] models, which may be made of various elements such as helium, carbon and iron. The Hamada–Salpeter models also use an equation of state due to Salpeter[4], which takes into account various forces between the particles of the degenerate matter.

Various models and the periods of their fundamental radial modes have been computed using the variational formula of Ledoux and Walraven[5]. Trial eigenfunctions of the form

$$\xi = r + \alpha r^2 + \beta r^3 + \gamma r^4 + \delta r^5 + k r^{256}$$

satisfying the boundary condition at the surface, were used. Newtonian mechanics was used.

To check the computer program, models obeying the Harrison–Wheeler–Wakano equation of state were computed and compared with Meltzer and Thorne's results (Figs. 1 and 3). There is fair agreement in both mass and radius. Periods were calculated for central densities ρc up to 10^{10} kg m^{-3}, and again there is fair agreement. The differences are such as may be attributed to the neglect of general relativity in the present work.

Similar computations were made for models composed of iron, of helium and of carbon, and obeying Salpeter's equation of state (Figs. 2 and 4). The masses and radii of the models (Fig. 2) agree precisely with Hamada and Salpeter's computations (which also omitted general relativistic effects). Helium becomes unstable against nuclear reactions when $\rho_c > 8 \times 10^{11}$, carbon when $\rho_c > 6 \times 10^{12}$, and iron when $\rho_c > 10^{12}$ kg m^{-3}; in each case, white dwarfs with higher central densities must contain cores of heavier or more neutron-rich elements[2].

A striking feature of Figs. 3 and 4 is that the oscillation period depends very little on composition, or on the equation of state used, until the sequence of models approaches instability. The rising curves of Figs. 3 and 4 may all be very nearly superposed. The point at which instability occurs is much more sensitive to such changes, and therefore the small periods for iron and carbon models of high density (Fig. 4) must not be taken very seriously until the calculations have been repeated with corrections for general relativity.

I thank Dr P. A. G. Scheuer and others in the Radio Astronomy Department, University of Cambridge, for invaluable help and advice; I am indebted to the Science Research Council for a maintenance grant.

Received April 29, 1968.

[1] Hewish, A., Bell, S. J., Pilkington, J. D. H., Scott, P. F., and Collins, R. A., *Nature*, **217**, 709 (1968), (Paper 1).
[2] Meltzer, D. W., and Thorne, K. S., *Ap. J.*, **145**, 514 (1966).
[3] Hamada, T., and Salpeter, E. E., *Ap. J.*, **134**, 683 (1961).
[4] Salpeter, E. E., *Ap. J.*, **134**, 669 (1961).
[5] Ledoux, P., and Walraven, T., *Handbuch der Physik*, **51**, 463 (1958).

31. Oscillations of Hamada–Salpeter White Dwarfs including General Relativistic Effects

by

J. SKILLING

Mullard Radio Astronomy Observatory,
Cavendish Laboratory,
University of Cambridge

FOLLOWING a recent communication[1], I have now computed radial oscillation periods for the fundamental modes of zero temperature white dwarfs obeying the Salpeter equation of state[2], using general relativistic, rather than Newtonian, dynamics. As before, a variational approach was used, which maximizes the period given by various trial eigenfunctions. The Tolman–Oppenheimer–Volkoff[3,4] equation of hydrostatic equilibrium

$$\frac{dp}{dr} = -\frac{G(\rho + p/c^2)(m + 4\pi r^3 p/c^2)}{r^2(1 - 2Gm/rc^2)}$$

and the equation for the mass enclosed by r

$$dm/dr = 4\pi r^2 \rho$$

were integrated numerically in parallel with several trial eigenfunctions of the form

$$\xi \propto r + \alpha r^2 + \beta r^3 + \gamma r^4 + \delta r^5 + kr^{256}$$

using the Gill modification of the Runge–Kutta procedure. In these formulae, r is the Schwarzschild radial co-ordinate, and m is the gravitationally effective mass enclosed by r. M, the surface value of m, is the Schwarzschild mass of the star, which is the mass "felt" by exterior orbiting bodies. The density ρ is not simply proportional to the baryon number density, as assumed by Salpeter[2], but contains a correction for the kinetic energy (mass) of the degenerate electrons.

For each trial eigenfunction, a period was obtained using the variational formula given by Harrison, Thorne, Wakano and Wheeler[5], which is equivalent to an earlier formula of Chandrasekhar[6]. The coefficient k of r^{256} was chosen to satisfy the surface boundary condition given by Bardeen, Thorne and Meltzer[7], the high exponent being used to satisfy the boundary condition without significantly affecting ξ through the main body of the star and therefore adversely affecting the variational accuracy. The computed periods were, in fact, only altered by less

than 1 part in 10^7 by ignoring the boundary condition and putting $k = 0$ instead.

Homogeneous stellar models composed of helium, of carbon, and of iron were used, with central densities ranging from 10^8 kg m^{-3} to about 3×10^{13} kg m^{-3}. Strictly, the denser models are unstable to nuclear reactions (see, for example, Salpeter[2]) and should have been given cores of different composition, but the periods obtained depended so little on composition that this addition to the program was not thought to be worthwhile. Table 1 lists the most accurate periods obtainable for a given star by varying α, β, γ and δ. These values are actually lower limits to the period of the true fundamental. The homologous eigenfunction $\xi \propto r$ gives periods differing from these values by 1 per cent or less, however, and it seems likely from intermediate results that the addition of terms up to r^5 will have reduced the error well below 1 per cent.

There have been three checks on the results. First, the general relativistic results fit smoothly onto the Newtonian results for stars of low density. Fig. 1, which shows the periods of general relativistic and of Newtonian stars, shows the fit clearly. Second, a selection of stars was re-calculated with shorter step lengths in the integration, and the results differed by under 1 part in 10^4. Third, the point of onset of dynamical instability coincides precisely with the peak of the mass-radius curve (see the appropriate entries in Table 1), as predicted by Wheeler's instability criterion[5].

As mentioned, composition makes remarkably little difference to the period of a star of given central density. The use of general relativity in the calculations causes an abrupt onset of instability around central densities of $2 \cdot 5 \times 10^{13}$ kg m^{-3}, however. At these densities, the electrons which provide the pressure are highly relativistic, so that $\gamma = (\rho/p + 1/c^2)dp/d\rho$ approaches $4/3$, which is a critical value for the overall stability of stellar models. When

Table 1. MASSES, RADII AND PERIODS OF MODELS OF DIFFERENT COMPOSITION

Central kg m⁻³	Helium-4			Carbon-12			Iron-56		
	M kg	R km	Period s	M kg	R km	Period s	M kg	R km	Period s
1·000 × 10⁸	2·791 × 10²⁹	1·569 × 10⁴	56·2	2·648 × 10²⁹	1·514 × 10⁴	54·7	1·826 × 10²⁹	1·259 × 10⁴	49·7
3·162 × 10⁸	4·783 × 10²⁹	1·304 × 10⁴	33·2	4·610 × 10²⁹	1·272 × 10⁴	32·6	3·382 × 10²⁹	1·099 × 10⁴	30·4
1·000 × 10⁹	7·760 × 10²⁹	1·073 × 10⁴	20·1	7·554 × 10²⁹	1·054 × 10⁴	19·9	5·767 × 10²⁹	9·328 × 10³	18·9
3·163 × 10⁹	1·166 × 10³⁰	8·731 × 10³	12·8	1·142 × 10³⁰	8·612 × 10³	12·7	8·958 × 10²⁹	7·743 × 10³	12·2
1·000 × 10¹⁰	1·598 × 10³⁰	7·004 × 10³	8·60	1·571 × 10³⁰	6·928 × 10³	8·54	1·256 × 10³⁰	6·295 × 10³	8·22
3·163 × 10¹⁰	2·000 × 10³⁰	5·531 × 10³	6·09	1·970 × 10³⁰	5·481 × 10³	6·05	1·596 × 10³⁰	5·016 × 10³	5·83
1·001 × 10¹¹	2·320 × 10³⁰	4·295 × 10³	4·50	2·288 × 10³⁰	4·261 × 10³	4·47	1·870 × 10³⁰	3·920 × 10³	4·31
3·165 × 10¹¹	2·543 × 10³⁰	3·276 × 10³	3·42	2·510 × 10³⁰	3·254 × 10³	3·42	2·064 × 10³⁰	3·006 × 10³	3·29
1·001 × 10¹²	2·680 × 10³⁰	2·455 × 10³	2·71	2·647 × 10³⁰	2·440 × 10³	2·70	2·184 × 10³⁰	2·262 × 10³	2·59
3·169 × 10¹²	2·756 × 10³⁰	1·808 × 10³	2·26	2·722 × 10³⁰	1·797 × 10³	2·25	2·252 × 10³⁰	1·671 × 10³	2·13
6·3 × 10¹²	2·780 × 10³⁰	1·49 × 10³	2·15*	2·746 × 10³⁰	1·48 × 10³	2·14*			
8·3 × 10¹²							2·279 × 10³⁰	1·28 × 10³	1·97*
1·003 × 10¹³	2·789 × 10³⁰	1·309 × 10³	2·24	2·755 × 10³⁰	1·303 × 10³	2·21	2·283 × 10³⁰	1·214 × 10³	2·00
2·80 × 10¹³	2·795 × 10³⁰	1·03 × 10³	∞						
2·35 × 10¹³				2·762 × 10³⁰	1·01 × 10³	∞			
2·89 × 10¹³							2·290 × 10³⁰	8·93 × 10²	∞
3·177 × 10¹³	2·794 × 10³⁰	9·345 × 10²	Unstable	2·761 × 10³⁰	9·299 × 10²	Unstable	2·290 × 10³⁰	8·686 × 10²	Unstable

* Denotes minimum period for a given composition. ∞ Denotes infinite period, and the onset of dynamical instability.

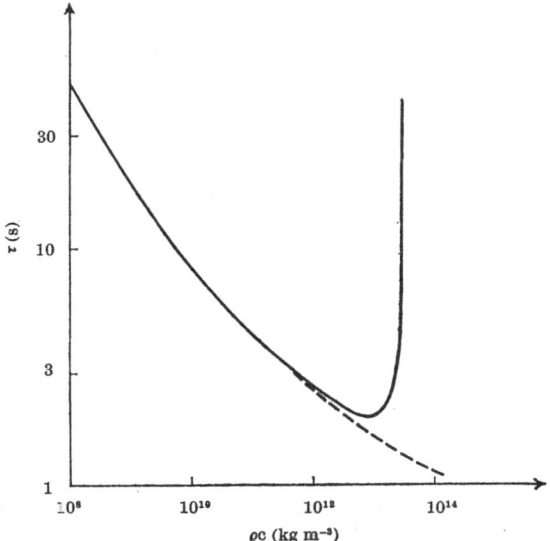

Fig. 1. Periods of Hamada-Salpeter models composed of iron. Continuous curve, general relativistic results; dashed curve, Newtonian results.

geometrical effects, which are governed by the parameter Gm/rc^2, become comparable with $\gamma - 4/3$, the star may become unstable, even though on Newtonian theory with flat space–time the star would be stable as $\gamma > 4/3$ everywhere. A more complete discussion of this point is given by Chandrasekhar[6]. Table 1 gives parameters of stars on the verge of dynamical instability (infinite period). The minimum periods obtainable for a given composition, which are essentially determined by the position of onset of instability, are also given in Table 1. From an inspection of these values, it seems it will not be possible to construct a homogeneous white dwarf, obeying the Salpeter equation of state, which would oscillate appre-

ciably faster than once every 2·0 s. This may be contrasted with the minimum of 7 s or so obtained by Meltzer and Thorne[8] for models obeying the Harrison–Wheeler–Wakano equation of state.

If pulsations of white dwarfs do cause the radio pulses discovered by Hewish et al.[9], these stars must be oscillating in overtones, and there must be some mechanism for exciting one particular overtone to the exclusion of the others. The source CP 0950 in particular, with a period of 0·253071 s (ref. 10), oscillates eight times faster than any stellar model discussed here. Of course, indefinitely high frequencies may be obtained by the use of sufficiently high overtones, for as the number of nodes in the eigenfunction increases, the overtone passes into a regime alternatively described as a standing sound wave in the star.

It is worth noting that the radii of these stars extend down to 900 km before dynamical instability occurs. This is rather smaller than the minimum radius for stable Harrison–Wheeler–Wakano models, and this may be of assistance in explaining the weak optical emission of pulsating radio sources—quite regardless of arguments about the possible connexion between stellar oscillations and the radio pulses.

I thank members of the radio astronomy group at Cambridge for their advice and encouragement.

Received May 13, 1968.

[1] Skilling, J., Nature, 218, 531 (1968), (Paper 30).
[2] Salpeter, E. E., Ap. J., 134, 669 (1961).
[3] Tolman, R. C., Relativity, Thermodynamics, and Cosmology, 242 (Clarendon Press, Oxford, 1934).
[4] Oppenheimer, J. R., and Volkoff, G., Phys. Rev., 55, 374 (1939).
[5] Harrison, B. K., Thorne, K. S., Wakano, M., and Wheeler, J. A., Gravitation Theory and Gravitational Collapse, 50, 63 (Univ. Chicago Press, 1965).
[6] Chandrasekhar, S., Ap. J., 140, 417 (1964).
[7] Bardeen, J. M., Thorne, K. S., and Meltzer, D. W., Ap. J., 145, 505 (1966).
[8] Meltzer, D. W., and Thorne, K. S., Ap. J., 145, 514 (1966).
[9] Hewish, A., Bell, S. J., Pilkington, J. D. H., Scott, P. F., and Collins, R. A., Nature, 217, 709 (1968), (Paper 1).
[10] Pilkington, J. D. H., Hewish, A., Bell, S. J., and Cole, T. W., Nature, 218, 126 (1968), (Paper 2).

32. Pulsation Periods of Rotating White Dwarf Models

by

J. P. OSTRIKER
J.-L. TASSOUL

Princeton University Observatory,
Princeton, New Jersey

THE recent discovery[1,2] of rapidly pulsating radio sources has renewed interest in the adiabatic oscillations of white dwarfs. While we believe that, for various reasons[3,4], the pulsars are not connected with pulsating white dwarfs, there is a certain intrinsic interest in the minimum oscillation period for an ordinary star, and in this work we will discuss the effect of rotation on the lowest modes of pulsation.

The periods of radial pulsation for completely degenerate, non-rotating white dwarfs obeying the Chandrasekhar equation of state[5] were first obtained by Sauvenier-Goffin[6], and were later corrected by Schatzman[7] for a minor numerical error. In the range of central densities where these models represent actual white dwarfs, the fundamental period of radial pulsation is always longer than 3 s. For more massive models, the equation of state has to be modified in order to include inverse beta decay[8-10] and general relativistic effects[11]. Using the Harrison–Wakano–Wheeler equation of state[12], Meltzer and Thorne[13] found that a non-rotating white dwarf cannot oscillate in the fundamental mode with a period less than 8 s. According to Faulkner and Gribbin[14], however, this result is due to an erroneous specification of the adiabatic exponent. When the Salpeter equation of state[9] and general relativity corrections are included in the treatment, the fundamental pulsation period for a white dwarf can be as low as 1·5 s (refs. 14 and 15). The fact that such models cannot oscillate in their fundamental mode with a period as short as the 0·25 s repetition rate found in CP 0950 (ref. 2) has led various authors[14-16] to suggest that, if pulsating stars are associated with "pulsars", we must be observing an excited overtone oscillation.

We will now show that, when rotation is taken into account, there exists a fundamental axisymmetric mode characterized by a period which can be as low as 0·25 s. The effects of rotation are manifold: (i) radial symmetry of the configuration is destroyed, so that all modes have some latitude dependence, (ii) the degeneracy in the frequency spectrum is removed and (iii) various physically distinct modes are coupled. This last property is particularly interesting. Indeed, rotation couples the two lowest modes which, in the limit of hydrostatic equilibrium, are independent, that is, the fundamental radial mode and the lowest axisymmetric Kelvin mode[17,18]. The theoretical expressions which were obtained by us in ref. 18 and which were solved numerically, are not very illuminating. The general effect of slow rotation on the two lowest axisymmetric modes can, however, be seen in the following approximate formulae

$$\sigma_R^2 \sim (3\bar\gamma - 4)\frac{|W|}{I}\left[1 + \alpha\left(\frac{5 - 3\bar\gamma}{3\bar\gamma - 4}\right)\frac{T}{|W|}\right] \qquad (1)$$

and

$$\sigma_Z^2 \sim \frac{4}{5}\frac{|W|}{I}\left[1 + \beta\frac{T}{|W|}\right] \qquad (2)$$

Here σ, $\bar\gamma$, T, W and I denote, respectively, the frequency of pulsation, the pressure weighted average of the adiabatic exponent, the kinetic energy, the potential energy and the moment of inertia of the rotating configuration. α and β are two positive constants of order unity[19,20]. For larger rotation, both modes depend on the adiabatic exponent and neither is purely radial. In very dense white dwarfs $\bar\gamma$ tends toward 4/3. Thus the rotational correction term in equation (1) can become significant even if $T/|W|\ll1$. There is an additional, negative, correction to σ_R^2 due to general relativity. It has been shown, in the absence of rotation, that relativistic effects initiate a dynamical instability for central densities of the order of $2\cdot3\times10^{10}$ g cm^{-3} (ref. 11). Because we are here considering models well below this limit and because, on the other hand, rotation is known to inhibit such an instability[21], we will

Table 2. AXISYMMETRIC PULSATION PERIODS OF NON-UNIFORMLY ROTATING WHITE DWARFS, IN A SEQUENCE OF FIXED MASS AND DECREASING ANGULAR MOMENTUM*

($M = 2\cdot26\ M_\odot$)

ϱ_c		J	$T/|W|$	P_R	P_Z
6·68	(8)	3·08 (50)	0·133	2·00	0·77
9·05	(8)	2·89 (50)	0·130	1·79	0·67
1·26	(9)	2·69 (50)	0·127	1·58	0·57
1·81	(9)	2·50 (50)	0·124	1·38	0·48
2·71	(9)	2·31 (50)	0·121	1·18	0·40
4·23	(9)	2·12 (50)	0·119	0·99	0·32
6·97	(9)	1·92 (50)	0·117	0·80	0·25
1·22	(10)	1·73 (50)	0·115	0·63	0·19

* The calculations are for $\mu_e = 2$. For other values of μ_e the Ps should be multiplied by $(2/\mu_e)^{1/2}$.

ignore the general relativistic correction. Finally, let us note that a vibrating "spheroidal" white dwarf will radiate gravitational waves. This may be considered a minor additional dissipative process.

Models of rotating white dwarfs obeying the Chandrasekhar equation of state have been constructed by James[22] and Ostriker et al.[23,24]. As shown by James, uniform rotation permits models only for a limited range of total angular momenta and masses; moreover, the upper mass, above which no equilibrium is possible, is not very much increased by a solid body rotation. No such restrictions exist when differential rotation is allowed[23,24]. Our present models were constructed according to the iterative method devised by Ostriker and Mark[25]. Approximate periods of the axisymmetric pulsations were computed by means of the second-order virial equations described in detail by us[18]. Table 1 summarizes computations made for uniformly rotating white dwarf models. In Table 2 we list the periods for differentially rotating white dwarfs of 2·26 $M_\odot$. For the latter models, the angular velocity of rotation is assumed to depend only on the distance from the axis of rotation; the angular momentum distribution is that of a uniformly rotating, homogeneous spheroid. In both cases, magnetic fields and dissipative effects were neglected. Having used the Chandrasekhar equation of state, we restricted ourselves to a range of central densities for which inverse beta-decay and general relativity are not yet very important. In both tables, the ratio $T/|W|$ is also given, for it provides an important clue concerning the stability of the models[26]. An integer in parentheses indicates the power of ten by which the corresponding entry should be multiplied. The central density ρ_c, the total angular momentum J and the two fundamental axisymmetric periods (P_R and P_Z) are given in c.g.s. units. In all our computations, μ_e, the atomic mass per unit electron, was taken equal to two.

It is apparent from Tables 1 and 2 that, as is well known, both periods strongly depend on the central density. Also, for given central density, the greater the total angular momentum, the smaller are the oscillation periods. From Fig. 1, comparing the various periods for different total angular momenta, we note that a uniform rotation does not greatly reduce the periods. The reason lies in the fact that, for a constant angular velocity, a white dwarf model is not able to sustain a very large kinetic energy T, when compared with the potential energy W. Even for $T/|W| \sim 10^{-3}$, however, P_Z may be as low as 0·3 s. This last figure can be reduced still further, if one constructs models very close to the limit beyond which equilibrium is no longer possible. As shown by

Table 1. AXISYMMETRIC PULSATION PERIODS OF UNIFORMLY ROTATING WHITE DWARFS FOR GIVEN ANGULAR MOMENTUM AND CENTRAL DENSITY*

ϱ_c (g cm^{-3})	$J = 1\cdot925$ (48)			ϱ_c (g cm^{-3})	$J = 3\cdot850$ (48)			ϱ_c (g cm^{-3})	$J = 7\cdot700$ (48)		
	$T/|W|$	P_R (s)	P_Z (s)		$T/|W|$	P_R (s)	P_Z (s)		$T/|W|$	P_R (s)	P_Z (s)
7·25 (8)	3·7 (−5)	2·84	0·93	7·19 (8)	1·5 (−4)	2·84	0·93	6·98 (8)	5·8 (−4)	2·85	0·95
1·10 (9)	4·0 (−5)	2·60	0·76	1·09 (9)	1·6 (−4)	2·60	0·77	8·47 (8)	6·0 (−4)	2·73	0·86
1·83 (9)	4·4 (−5)	2·35	0·60	1·80 (9)	1·8 (−4)	2·36	0·60	1·05 (9)	6·3 (−4)	2·61	0·78
3·50 (9)	5·2 (−5)	2·10	0·43	3·41 (9)	2·1 (−4)	2·10	0·44	1·32 (9)	6·6 (−4)	2·49	0·70
3·85 (9)	5·3 (−5)	2·06	0·41	5·10 (9)	2·3 (−4)	1·97	0·36	1·70 (9)	6·9 (−4)	2·36	0·62
4·25 (9)	5·5 (−5)	2·03	0·39	5·70 (9)	2·4 (−4)	1·93	0·34	2·25 (9)	7·4 (−4)	2·24	0·54
4·72 (9)	5·6 (−5)	2·00	0·37	6·42 (9)	2·4 (−4)	1·90	0·32	3·11 (9)	8·0 (−4)	2·10	0·46
5·27 (9)	5·8 (−5)	1·97	0·36	7·26 (9)	2·5 (−4)	1·87	0·30	4·50 (9)	8·8 (−4)	1·97	0·38
5·92 (9)	6·0 (−5)	1·94	0·34	8·28 (9)	2·6 (−4)	1·84	0·28	6·95 (9)	9·9 (−4)	1·83	0·31

* The calculations are for $\mu_e = 2$. For other values of μ the Ps should be multiplied by $(2/\mu_e)^{1/2}$.

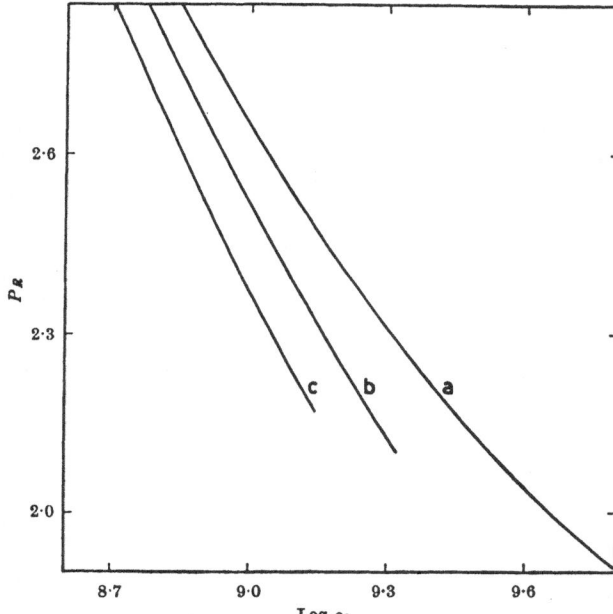

Fig. 1. Pulsation periods of uniformly rotating white dwarfs. The curves labelled *a*, *b* and *c* correspond, respectively, to $J = 3.85$ (48), 2.31 (49) and 3.85 (49). The zonal period P_Z is not shown because the effect of rotation is too small to be apparent on this scale.

all models given in Table 2 are secularly and ordinarily stable (in the Poincaré sense; our unpublished work).

To sum up, a rotational motion couples the two lowest axisymmetric modes of pulsation of a white dwarf obeying the Chandrasekhar equation of state. Neither of these modes is purely radial, and one of them may have a period as low as 0·25 s for a central density less than 10^{10} g cm⁻³. Very recently, Durney *et al.*[26] presented calculations indicating that the P_R modes of differentially rotating stars may also oscillate with a period as short as 0·4 s for a central density of the order of 10^{11}.

This work was supported, in part, by grants from the US Air Force Office of Scientific Research and the US National Science Foundation.

Received July 19, 1968.

[1] Hewish, A., Bell, S. J., Pilkington, J. D. H., Scott, P. F., and Collins, R. A., *Nature*, 217, 709 (1968), (Paper 1).
[2] Pilkington, J. D. H., Hewish, A., Bell, S. J., and Cole, T. W., *Nature*, 218, 126 (1968), (Paper 2).
[3] Ostriker, J. P., *Nature*, 217, 1227 (1968), (Paper 39).
[4] Gold, T., *Nature*, 218, 731 (1968), (Paper 40).
[5] Chandrasekhar, S., *Introduction to the Study of Stellar Structure* (University of Chicago Press, 1939).
[6] Sauvenier-Goffin, E., *Mém. Soc. Roy. Sci. Liège*, 10, 1 (1950).
[7] Schatzman, E., *Ann. Astrophys.*, 24, 237 (1961).
[8] Schatzman, E., *Astron. Zhurnal*, 33, 800 (1956).
[9] Salpeter, E. E., *Ap. J.*, 134, 669 (1961).
[10] Hamada, T., and Salpeter, E. E., *Ap. J.*, 134, 683 (1961).
[11] Chandrasekhar, S., and Tooper, R. F., *Ap. J.*, 139, 1396 (1964).
[12] Harrison, B. K., Thorne, K. S., Wakano, M., and Wheeler, J. A., *Gravitation Theory and Gravitational Collapse* (University of Chicago Press, 1965).
[13] Meltzer, D. W., and Thorne, K. S., *Ap. J.*, 145, 514 (1966).
[14] Faulkner, J., and Gribbin, J. R., *Nature*, 218, 734 (1968), (Paper 28).
[15] Skilling, J., *Nature*, 218, 923 (1968), (Paper 30).
[16] Thorne, K. S., and Ipser, J. R., *Ap. J.*, 152, L71 (1968).
[17] Chandrasekhar, S., and Lebovitz, N. R., *Ap. J.*, 135, 248 (1962).
[18] Tassoul, J. L., and Ostriker, J. P., *Ap. J.* (in the press).
[19] Ledoux, P., *Ap. J.*, 102, 143 (1945).
[20] Chandrasekhar, S., and Lebovitz, N. R., *Ap. J.*, 136, 1082 (1962).
[21] Fowler, W. A., *Ap. J.*, 144, 180 (1966).
[22] James, R. A., *Ap. J.*, 140, 552 (1964).
[23] Ostriker, J. P., Bodenheimer, P., and Lynden-Bell, D., *Phys. Rev. Lett.*, 17, 816 (1966).
[24] Ostriker, J. P., and Bodenheimer, P., *Ap. J.*, 151, 1089 (1968).
[25] Ostriker, J. P., and Mark, J. W. -K., *Ap. J.*, 151, 1075 (1968).
[26] Durney, B. R., Faulkner, J., Gribbin, J. R., and Roxburgh, I. W., *Nature*, 219, 20 (1968), (Paper 29).

James[22], all uniformly rotating white dwarfs are secularly stable. The situation is very different for non-uniformly rotating white dwarfs above the Chandrasekhar critical mass. Differential rotation allows much larger values of $T/|W|$ and has, consequently, a much larger effect on the periods. For $\rho_c < 10^{10}$ g cm⁻³, the zonal period may be as low as 0·25 s. It is worth noting that, for a given mass (not density), rotation increases the periods. This can easily be understood from equations (1) and (2), for large rotation considerably reduces the potential energy and increases the moment of inertia. Finally, let us note that

33. Model for Pulsed Radio Sources

by
WERNER ISRAEL
Dublin Institute for Advanced Studies,
Dublin.

INTENSIVE study of the remarkable pulsed radio sources announced by the Mullard group[1] is just beginning. It is nevertheless already possible, independently of any particular model, to draw sharp and significant conclusions from the observational data now to hand.

Radiation from all four sources is strongly polarized[2], and so is definitely of synchrotron origin. If E (in MeV) represents a mean energy for electrons contributing significantly to the synchrotron flux, then the frequency at which the flux is a maximum can be roughly estimated as $\nu_{max} = 5 \times 10^6 HE^2$ s⁻¹ (H in gauss). The source CP 1919 has been observed at five frequencies[3] between 81·5 MHz and 2,695 MHz, and shows a steeply declining spectrum. It follows that $\nu_{max} \lesssim 10^8$ s⁻¹. On the other hand, the half-life of the synchrotron electrons, $\tau_{1/2} \approx (2\cdot6 \times 10^8)/H^2 E$ s, cannot be larger than the observed pulse-width of 35 ms. From these two inequalities we derive $H \gtrsim 10^6$ gauss, $E \lesssim 4 \times 10^{-3}$ MeV. Thus the effective synchrotron electrons have velocity components normal to the field lines not exceeding 0·05 c.

The total radio emission is roughly

$$L \approx 2 \times 10^{-22} H(\Delta\nu)\, n \text{ erg s}^{-1}$$

where n is the total number of effective synchrotron electrons and $\Delta\nu \approx 5 \times 10^8$ s⁻¹ is an effective bandwidth. Inserting the observed value $L \approx 10^{29}$ erg s⁻¹ yields $n \lesssim 10^{36}$. It follows that the total mass of plasma involved in the synchrotron mechanism cannot exceed 10^{14} g, if we assume that about 1 per cent of the electrons are effective synchrotron emitters. This leads to the idea that all the radio emission originates from a very light shell. Lyne and Smith[2] have advanced arguments, based on the coherence of the polarized signals, for believing that at least one linear dimension of the source is less than 10 cm.

Most remarkable of all is the unprecedented sharpness of the pulses. This emerges with particular clarity from the polarized recordings of Lyne and Smith[2]. It seems virtually impossible to reconcile such a pulse shape with any continuous oscillation of the source. The observations rather suggest some sort of "clockwork mechanism" which triggers off short impulses (≈ 30 ms) separated by comparatively long intervals (≈ 1 s) of quiescence. The precise regularity of the signals attests to the delicate balance of this clockwork; its astonishing stability in the face of the violent fluctuations in pulse amplitude must mean that it is energetically decoupled from the radio source, or (what is really the same thing) that it is controlled by an energetic process of a much higher order of magnitude.

The model I propose is that a pulsar is a neutron star on the edge of gravitational instability. The pulsing mechanism is assumed to be a constant repetition, in practically identical conditions, of the following cycle of events. A very light additional layer of matter settles on the surface of the star, initiating an instability, and the star begins to collapse. The surface layers will usually retain a small reserve of nuclear energy. Even a small outward flow of radiation will partially relieve the core of the weight of the overlying layers, so stability is restored. The core snaps back, throwing off a thin outer layer with less than escape velocity, but sufficient to raise some of the material to a maximum height $h \approx 10^9$ cm. After an interval of $\pi(h^3/2GM)^{1/2} \approx 1$ s, this material will all have trickled back onto the surface, triggering off the next instability. The "clockwork" in this model is thus a short-period "astronomical sand-glass".

Let us make some rough numerical estimates. For convenience, we adopt the round numbers $M = 3 \times 10^{33}$ g, $R = 10^6$ cm for the critical mass and radius of the neutron star. The degenerate core is surrounded by a thin, hydrogen-rich, non-degenerate envelope, the mass of which can reasonably be estimated at $m \approx 10^{12}$–10^{13} g for a cold neutron star. For the thermal energy generated by incipient collapse we assume a value $\approx mc^2 \approx 10^{34}$ ergs. The resulting energy density is of order 10^{16} ergs cm^{-3}, corresponding to X-ray phonons. The degenerate interior of the star has a very high thermal conductivity, and the X-ray pulse is transmitted with the speed of sound ($\approx c/\sqrt{3}$) to the surface. Here it blows off the cool non-degenerate shell with moderate relativistic velocity, rapidly converting it into plasma at $\approx 10^7$ °K. The shell material is highly transparent to X-rays in these conditions, and the X-ray pulse will pass freely into space within the first millisecond.

Thus within the first few milliseconds of the initial shock considerable cooling will have taken place, and synchrotron radiation will be the main source of emission of the expanding shell. As we have seen, the synchrotron electrons must decay within 35 ms. At this stage, the shell's radius is still less than 10^7 cm ≈ 0.3 light-ms. On the present model, the observed pulse-width is therefore determined by the synchrotron decay-time and not by the size of the source. The fact that CP 0950, with a very small pulse-width, also appears to have an unusually strong magnetic field tends to support this conclusion.

The maximum height $\approx 10^9$ cm to which the ejected material rises is still well within the radius of gravitational influence of the star ($\approx 10^{14}$ cm). Nevertheless, small losses in the number of baryons are inevitable, and such a delicately balanced mechanism would soon run down if left to itself. It is here that accretion plays a decisive part. A star with velocity 10^6 cm s^{-1} would pick up $\approx 10^{11}$ g s^{-1} from the interstellar medium. In the case of a pulsar, most of this is dispersed by the X-ray blasts, but only a very small fraction (perhaps a millionth) is needed to maintain the balance. In the long run, the scale of the pulsar mechanism must be regulated by the supply of accreted material. It may therefore be significant that CP 1919, which is in the plane of the galaxy where the interstellar density is highest, is intrinsically the most powerful of the four known sources. It is more than six times as powerful as CP 0950 and CP 1133,

which lie a considerable distance off the plane. Of course, this argument ignores possible differences in the velocities and magnetic fields.

On the basis of this model a number of predictions concerning the pulsars, both qualitative and quantitative, can be made. Most interesting is the possibility of detecting them as X-ray sources. The most favourable case, because of its proximity, is CP 0950 for which the observed flux should be of the order of 10^{-8} ergs cm^{-2} s^{-1}. There is, in fact, a possibility that CP 0950 has already been observed as an X-ray source. In 1966, the Naval Research Laboratory group[4] reported a source (Leo X-1) at $\alpha = 9^{\rm h}35^{\rm m}$, $\delta = +8°\cdot6$. The source was observed on only one scan (it has apparently not been seen since) and there are pitch uncertainties amounting to $20^{\rm m}$ in α and $0°\cdot7$ in δ. The position of CP 0950 is $\alpha = 9^{\rm h}50^{\rm m}$, $\delta = 8° \ 10'$. It is thus possible that CP 0950 and Leo X-1 (if its existence can be confirmed) are one and the same object.

Radiation at optical wavelengths would be completely negligible on the present model. It is therefore unlikely that CP 1919 has any connexion with the blue star tentatively identified with it by Ryle and Bailey[5].

How does a pulsar arrive in such a delicately poised and apparently implausible condition? We suggest that, on the contrary, the pulsar condition is a natural stage in the evolution of many (and perhaps most) supernovae remnants. If, after the explosion and fragmentation of the supernova, one of the fragments is still somewhat above the critical mass, it will undergo violent gravitational oscillations, ejecting relativistic particles and clouds of gas. (That such a condition must terminate in a "Schwarzschild singularity" is a widespread belief which actually has little to support it, particularly when departures from spherical symmetry are taken into account[6].) The amplitude of the oscillations will gradually die down as the mass declines toward the critical mass (the Crab nebula is probably an example of the final part of this stage). Eventually the star evolves into the state described here.

What is the lifetime of a pulsar? Accretion will supply enough baryons to keep the star at or just above the critical mass indefinitely. But the energy output (at least on the scale considered here) must come from the remnants of the internal nuclear resources. In fact, the requirements are very modest: an internal energy of 10^{48} ergs will maintain the process on the scale considered here for 10^6 yr. The actual resources are likely to be some orders of magnitude greater than this.

From the point of view of its existence as a radio source, a much more stringent restriction on the lifetime of a pulsar is imposed by magnetic decay. At an interior point where the density has risen to 10^{11} g cm^{-3} (just before neutronization supervenes) we may take $\sigma \approx 10^{23}$ s^{-1} as a very rough figure for the conductivity of the degenerate electron gas[7]. This gives a half-life for the magnetic field $t_{1/2} \approx \sigma R^2/c^2 \approx 3 \times 10^6$ yr, in satisfactory agreement with other estimates[8].

If these considerations are correct, then there must exist in the galaxy a comparatively large number of radio-quiet pulsars which remain fairly powerful X-ray sources. The reason why such objects should appear to show a preference for the galactic plane has already been explained.

I thank Dr I. Khan for stimulating discussions.

Received April 29, 1968.

[1] Hewish, A., Bell, S. J., Pilkington, J. D. H., Scott, P. F., and Collins, R. A., *Nature*, **217**, 709 (1968), (Paper 1).

[2] Lyne, A. G., and Smith, F. G., *Nature*, **218**, 124 (1968), (Paper 7).

[3] Davies, J. G., Horton, P. W., Lyne, A. G., Rickett, B. J., and Smith, F. G., *Nature*, **217**, 910 (1968), (Paper 6). Moffet, A. T., and Ekers, R. D., *Nature*, **218**, 227 (1968), (Paper 9).

[4] Byram, E. T., Chubb, T. A., and Friedman, H. L., *Science*, **148**, 152 (1966).

[5] Ryle, M., and Bailey, J. A., *Nature*, **217**, 907 (1968), (Paper 19).

[6] Israel, W., *Nature*, **216**, 148 (1967); ibid., **216**, 312 (1967); *Commun. Math. Phys.*, **8**, 245 (1968).

[7] Kothari, D. S., *Phil. Mag.*, **13**, 361 (1932).

[8] Pilkington, J. D. H., Hewish, A., Bell, S. J., and Cole, T. W., *Nature*, **218**, 126 (1968), (Paper 2). Hoyle, F., and Narlikar, J., *Nature*, **218**, 123 (1968), (Paper 27).

34. Pulsar Condition in White Dwarf Stars

by

WERNER ISRAEL

Dublin Institute for Advanced Studies,
Dublin

ACCORDING to the recently proposed "astronomical sandglass model" of the pulsed radio sources[1] a pulsar is a neutron star at the critical mass limit for gravitational instability. Any loss of stability resulting in incipient collapse is quickly restored by the remaining nuclear resources, which, it was assumed, can still generate thermal pulses of about 10^{34} ergs. Because the radius of a critical neutron star is 10^6 cm, the resulting energy density is of order 10^{16} erg cm^{-3}, corresponding to X-ray phonons. The degenerate interior of the star has a very high thermal conductivity and the X-ray pulse passes rapidly to the surface, where it blows off the thin ($\sim 10^{12}$ g) non-degenerate outer layer with less than escape velocity. After about a second this ejected material has all trickled back to the surface, triggering off the next instability.

It was conjectured that neutron star pulsars could have evolved from Type II supernova remnants, the masses of which remain somewhat above critical mass, by violent gravitational oscillations with emission of gas and relativistic baryons. In these cases, the critical mass limit is approached from above. The process of accretion offers a mechanism by which an underluminous white dwarf star might approach the critical mass from below. In this communication I consider the possibility that the pulsar condition might be characteristic of a sizable proportion of those Population II stars the evolution of which can be significantly affected by accretion. This hypothesis is not unreasonable for certain classes of stars, for the rate of accretion for a star of mass M passing with speed v through a medium of density ρ is

$$\frac{\mathrm{d}\,(M/M_\odot)}{\mathrm{d}\,(t/10^9\ \mathrm{yr})} \approx 0 \cdot 02 \left(\frac{M}{M_\odot} \right)^2 \frac{\rho}{(10^{-23}\ \mathrm{g\ cm^{-3}})}\ (v/1\ \mathrm{km\ s^{-1}})^{-3}.$$

In close binaries, where ρ may exceed the normal interstellar density by several orders of magnitude, and in the cores of galaxies where stars move slowly with respect to the medium, accretion can bring a weakly radiating star up to the critical mass in a time small compared with the age of the universe.

Consider now a white dwarf star which has reached the verge of gravitational instability in this manner. For the critical mass and radius of the degenerate core we can adopt the round numbers $M = 3 \times 10^{33}$ g, $R = 2 \times 10^8$ cm. From the analysis of stellar envelopes given by Chandrasekhar[2] we deduce that the core is surrounded by a non-degenerate shell of mass $\sim 10^{23}\ [L(1-X)]^{5/7}$ g and having an effective temperature $10^5\ L^{1/4}$ °K, where L is the luminosity in terms of the luminosity of the Sun taken as unity, and X the proportion by weight of hydrogen in the envelope. In normal conditions in the interior of a white dwarf, the remaining nuclear resources will suffice to maintain a continuous emission of up to $10^{-3} - 10^{-1}\ L_\odot$. If the radiation pressure thus produced can limit the inflow of material to $\sim 10^9$ to 10^{11} g s^{-1}, the star can exist in a steady "smouldering" condition at the critical mass, radiating away exactly as much mass as it accretes.

If, on the other hand, the rate of accretion cannot be kept under control by the available radiation pressure, a stationary state cannot be maintained. It then seems plausible that the star will pass over into a pulsar condition, gaining temporary reprieves from instability by lifting off its non-degenerate outer shell (now highly distended) with a small velocity by a series of pulses, which also sweep away much of the accreting material. The pulses are generated in the deeper layers by momentary instabilities caused by descent of the shell onto the surface. If we assume a pulse energy of 10^{35} ergs, the initial pulse (before passage through the envelope) will be in the form of soft X-rays.

There are now two alternatives. First, if the shell is opaque to X-rays, the pulses, in diffusing through it, will be degraded to near optical frequencies and the pulsed character of the emission will be visible as a smoothed-out light variation. The period is then simply the time required for a pulse to traverse the envelope (after which the full weight of the shell again bears down on the surface). A detailed analysis is difficult because of the sensitive temperature-dependence of the coefficient of opacity, but rough estimates suggest that it might be possible to account for the observed 71 s variation of Nova DQ Herculis[3] in this way. Other explanations for this variation are, of course, possible, but they would probably not account for the rapid flickering, which can here be ascribed to relaxation oscillations of a star close to neutral equilibrium. The second alternative—an envelope transparent to X-rays—may arise in the case of very close binaries, where tidal and rotational effects may concentrate the shell into an equatorial disk. X-ray emission should be directly observable in this case. How this relates to observed X-ray sources is a question which need not be pursued further, for Shklovsky[4] has inferred a model of the general type I have described directly from the observed properties of Sco X-1. The chief new aspect raised here is the possibility of a pulsed character for the X-ray emission.

We can now easily imagine circumstances (for example, a hydrodynamical instability) under which the shock mechanism envisaged for controlling the accretion breaks down. A sudden increase in the rate of accretion, say to 10^{17}–10^{18} g s^{-1}, would appreciably affect the mass of the envelope in the space of a day. The strength of the pulses and the internal temperatures generated would progressively increase until the envelope is blown off with escape velocity. The relief thus afforded the star would last only until a new envelope is formed—a matter of hours. The end result must be an ignition of powerful nuclear reactions in the deeper interior, with expulsion of perhaps 10^{28}–10^{30} g of material.

This picture of the initial stages of a nova outburst reproduces many of the observed features. In particular, it explains why several shells are ejected, and why the first shell is often overtaken by subsequent shells[5,6]. It also accounts for the concentration of novae towards the galactic plane, and for the fact that recurrent novae seem always to occur in close binary pairs[3] with an inverse correlation between strength of outburst and period of recurrence (compare with ref. 7). It has been estimated[8] that a recurrent nova loses about 10^{28} g in an outburst; if we take 10^{19} g s^{-1} as a reasonable figure for the accretion while the mass remains subcritical, then the time required to return to critical mass is 30 yr, which is about the observed period. (After many such outbursts, during which the supply of ^{3}He and heavy elements in the interior becomes steadily depleted, the star may terminate this phase of its career with a Type I supernova explosion, ending as a neutron star with subcritical mass.)

In these respects the model is very satisfactory. At the same time, it could be instantly disproved by a single well established instance of a recurrent nova which is not a member of a close binary pair, or of a nova mass which differs appreciably from the critical mass. Here, the recurrent nova T Coronae Borealis, which forms a spectroscopic binary, would seem to be the most clear-cut test case. Kraft[3] has obtained a minimum mass of $2 \cdot 1 \, M_\odot$ for the blue component, but Schatzman[6] has conjectured that this component may itself be a close binary, so the question seems to be still open.

Finally, consider briefly the implications of the hypothesis for the large-scale dynamics of galaxies. It is known from infrared observations[9] that there are $\sim 10^8$ stars within 20 pc of the centre of our galaxy which are invisible to us at optical and X-ray wavelengths. If the present ideas are correct many of these will be at the edge of gravitational instability, in a smouldering or a pulsar condition, with effective temperatures between 10^5 and 10^7 °K. Their radiated output will peak in the soft X-ray range and may amount to as much as 10^{-3} of the optical output of a centrally condensed galaxy. The integrated X-ray output of all galaxies should therefore be easily detectable, and it may be conjectured that this is the origin of the observed diffuse X-ray background. According to this picture, the core of a galaxy is periodically in a highly inflammable condition. The spark that ignites it could be (as suggested by Burbidge[10]) a single supernova explosion. The resulting shock will set off 10^7 or 10^8 novae

(perhaps 0·1 per cent of these will explode as Type I supernovae). After $\sim 10^6$ yr of accretion, the stage will be set for the next explosion.

Energies obtainable in this way ($\lesssim 10^{55}$ ergs) are still well short of those required to explain the quasi-stellar sources. Here it must be remembered that the very first generation of stars in the core of a galaxy were probably Population I objects which ended their lives as Type II supernovae, conceivably in a chain reaction[10]. In these conditions, a large fraction of the angular momentum of the innermost central stars would have been transferred to the radiation and to relativistic electrons, resulting in large magnetic fields, and an oscillatory collapse of the central portions of the galaxy. Magnetic forces will create a strong flow of electrons towards the centre, resulting in a charge excess and ejection of relativistic plasma along the axis, in accordance with the observed jetlike appearance.

Received May 10, 1968.

[1] Israel, W., Nature, **218**, 755 (1968), (Paper 33).

[2] Chandrasekhar, S., Introduction to the Study of Stellar Structure (Dover, New York, 1957).

[3] Kraft, R. P., Adv. Astro. Astrophys., **3**, 43 (1963).

[4] Shklovsky, I. S., Astrophys. J. Lett., **148**, L1 (1967). Prendergast, K. H., and Burbidge, G. R., Astrophys. J. Lett., **151**, L83 (1968).

[5] Kopal, Z., in Vistas in Astronomy (edit. by Beer, A.), **2**, 1491 (Pergamon Press, London, 1956).

[6] Schatzman, E., in Stellar Structure (edit. by Aller, L. H., and McLaughlin, D. B.), 327 (University of Chicago Press, Chicago, 1963).

[7] Huang, S. S., and Struve, O., Occasional Notes, Roy. Astro. Soc., **19**, 161 (1957).

[8] Payne-Gaposhkin, C., The Galactic Novae (North-Holland Pub. Co., Amsterdam, 1957).

[9] Becklin, E. E., and Neugebauer, G., Astrophys. J., **151**, 145 (1968).

[10] Burbidge, G. R., Nature, **190**, 1053 (1961).

35. Crystallization and Torsional Oscillations of Superdense Stars

by

M. A. RUDERMAN*
Department of Physics,
Imperial College, London

THE nuclei of a piece of iron compressed to a density of 10^8 g cm^{-3} will arrange themselves into a body centred cubic lattice with a melting temperature near 2×10^8 °K, corresponding to a thermal energy about 1 per cent of the coulomb repulsion between neighbouring nuclei[1]. For matter near the end point of thermonuclear evolution compressed to still higher densities the dominant nuclear species shift towards very neutron rich nuclei, with Z between 30 and 50, which arrange themselves into crystal lattices with an even greater melting temperature. Crystallization among nuclei can occur up to densities close to that of conventional nuclear matter where those protons that remain cluster into very neutron-rich nuclei which are surrounded by and exchange neutrons with an ambient degenerate neutron sea. Because canonical neutron star interiors are estimated to cool to an average temperature below 5×10^8 °K in less than 10^3 years[2-4], the outer regions of such stars should be solid. Beginning a few metres below the surface and extending almost to where nuclear densities are reached, neutron star matter will contain a crystal lattice that can support transverse shear waves. For very light neutron stars the lattice can fill the entire stellar volume. Because the restoring forces which oppose any shear strain in such lattices are caused

solely by coulomb interactions, they are much weaker than those forces which oppose density increases (electron or neutron degeneracy pressure, and repulsive neutron–neutron forces). They are also very much smaller than the equally strong gravitational force which acts to restore any deviations from a spherical shape of a neutron star. Consequently the lowest torsional modes of a neutron star, which preserve both stellar density and shape, have periods very much longer than the usual ones which preserve neither density nor shape. The torsional oscillations occur at frequencies which are proportional to the stellar radius but not sensitive to its mass and density distribution. For a radius $R \sim 20$ km the lowest torsional mode period is around 0·1 s, almost two orders of magnitude longer than the periods of non-torsional vibration of most neutron stars[5]. The similarity between the calculated torsional mode frequencies and those of the pulsed radio sources[6] suggests that it may be worth considering a possible relationship between the two, especially because the state of matter in the non-solid core of heavier neutron stars is such as to suppress the usual damping mechanisms for these oscillations.

When matter is compressed to densities exceeding 10^8 g cm^{-3} the Fermi energy of the degenerate electrons (> 1 MeV) greatly exceeds their coulomb interactions with

* On leave from New York University.

nuclei: the electron wave functions are only slightly distorted from plane waves. Consequently, except for the maintenance of overall electrical neutrality, they are not very effective in screening the coulomb charge of the nuclei embedded in them and the electrons can be approximated by an inert uniform sea of negative charge. (For a nuclear charge Z, the ratio of screening radius to internuclear separation is $\sim 4Z^{-1/3}$ for all densities as long as the electrons are relativistically degenerate.) In their lowest energy state the nuclei form a body centred cubic lattice[7,8] that can support two transverse shear waves, the long wavelength velocity of which depends only slightly on direction. Extensive numerical calculations of the crystal dispersion relations of a coulomb lattice have been performed by Clark[9] and others. The velocity of a low frequency transverse shear wave c_s in the [110] direction is

$$c_s = (0\cdot21)\,\frac{Zen_z^{2/3}}{\rho^{1/2}} \tag{1}$$

where n_z is the number density of nuclei with charge Ze. The inclusion of screening, which suppresses the interaction with distant neighbours, qualitatively changes the nature of long wavelength longitudinal compressional waves; but it has a small effect on the transverse shear waves and equation (1) is probably an overestimate. The crystal lattice can support long wavelength shear waves of this velocity until it melts at a temperature T_m given by

$$k\,T_m = Z^2e^2n_z^{1/3}\,\delta \tag{2}$$

with k the Boltzmann constant and δ, a dimensionless constant, is taken to be 1/50. Equation (2) is the only one allowed for a coulomb lattice by dimensional considerations. The coefficient δ is predicted to be 1/20 by Lindemann's rule[10,11]. A computer experiment on the analogue of a liquid–solid phase transition among thirty-two charged particles suggests $\delta \sim 1/80$ (ref. 12).

hemispheres rotate oppositely. The displacement s of any point (r,θ) within the sphere is given by

$$\mathbf{s} = \hat{\varphi}\,\sin 2\theta\,j_2\left(2\cdot7\,\frac{r}{R}\right)e^{2\pi i\nu t} \tag{3}$$

with frequency

$$\nu = \left(\frac{0\cdot43\,c_s}{R}\right) \sim \frac{2\times10^7}{R} \tag{4}$$

(Alfvén waves of this frequency will be emitted along any magnetic field lines which penetrate the oscillating stellar surface.)

The calculated radius R for any given neutron star mass depends on the detailed equation of state of dense neutron matter but not in a very sensitive way. The HWW equation of state[14] predicts for a stellar mass $M = 0\cdot26\,M_\odot$, a central density $\rho_c = 10^{14}$ g cm^{-3} and $R = 36$ km. The corresponding period of torsional oscillation is 0·2 s. Tsuruta and Cameron used equations of state which more explicitly incorporate knowledge of nuclear forces and calculated for $M = 0\cdot2\,M_\odot$ a radius of 18 km, and ρ_c of either $2\cdot4\times10^{14}$ g cm^{-3} or $3\cdot6\times10^{14}$ g cm^{-3} depending on detailed assumptions about the nucleon–nucleon interaction[17]. In either case the predicted torsional period is 0·1 s. The largest masses which give stable neutron stars ($M \lesssim 2\,M_\odot$) have calculated radii of about 10 km.

With the more realistic TC equation of state none except the very lightest neutron stars ($M < 0\cdot2\,M_\odot$) should have crystal lattice structure extending to the centre of the star. For $M \sim 2\,M_\odot$ the lattice should disappear after penetrating only about 0·5 km below the stellar surface. If oscillations of the sort described by equation (1) were restricted only to a freely oscillating outer solid mantle, the characteristic periods would be somewhat greater. The inner nucleon fluid of the core remains, however, in contact with the oscillating mantle and could exert a viscous drag on its motion. A weakly interacting normal

Table 1. COMPUTED MELTING TEMPERATURE AND TRANSVERSE SHEAR WAVE VELOCITIES FOR DENSITIES UP TO NUCLEAR DENSITY

ϱ (g cm^{-3})	10^7	10^8	10^9	$1\cdot7\times10^{10}$	$3\cdot2\times10^{11}$	$1\cdot6\times10^{12}$	$4\cdot5\times10^{12}$	$1\cdot0\times10^{14}$	2×10^{14}
Z	26	26	28	31	39	42	48	20	1
n_z(cm^{-3})	$1\cdot1\times10^{29}$	$1\cdot1\times10^{30}$	10^{31}	$1\cdot3\times10^{32}$	$1\cdot6\times10^{33}$	$2\cdot4\times10^{33}$	$3\cdot9\times10^{33}$	$2\cdot5\times10^{34}$	10^{36}
ϱ_n (g cm^{-3})	0	0	0	0	10^8	$1\cdot0\times10^{12}$	$3\cdot3\times10^{12}$	$1\cdot0\times10^{14}$	3×10^{14}
T_m (°K)	1×10^8	2×10^8	6×10^8	2×10^9	6×10^9	7×10^9	1×10^{10}	5×10^9	—
c_s (cm/s)	2×10^7	3×10^7	4×10^7	6×10^7	9×10^7	6×10^7	5×10^7	2×10^7	—

In the regime $\varrho = 10^7 - 10^8$ g cm^{-3} all protons are assumed bound into ^{56}Fe nuclei. The n_z and Z at $\varrho = 10^9$ g cm^{-3} are based upon an extrapolation of calculations by Tsuruta and Cameron[13]. Between $\varrho = 1\cdot7\times10^{10}$ and $4\cdot5\times10^{12}$ g cm^{-3} they are those of Harrison, Thorne, Wakano and Wheeler[14]. At $\varrho = 1\cdot0\times10^{14}$ g cm^{-3} the data are obtained by minimizing the energy of neutrons, protons and electrons using Weiss's calculations[15] of the energy of nuclear matter for various proton-neutron ratios, together with the assumption that the surface energy is proportional to the difference in energy between nucleons within a nucleus and in the surrounding neutron sea. Proton clustering into nuclei and hence lattice formation are found to disappear before canonical nuclear densities ($\varrho \sim 3\cdot6\times10^{14}$ g cm^{-3}) are reached. The density ϱ_n is that of unbound neutrons.

Calculated shear wave velocities and melting temperatures for very dense matter are given as a function of density in Table 1. At densities approaching those of normal nuclear matter the protons that are present in the neutron sea no longer cluster to form neutron rich ($A \sim 3\cdot5\,Z$) nuclei because neutron matter becomes too incompressible. Wolf[16] has estimated that nuclei will disappear before $\rho_n \sim 3\times10^{14}$ g cm^{-3}. My calculations, which lead to the data at $\rho = 1\cdot0\times10^{14}$ g cm^{-3}, also suggest a disappearance of nuclei near $\rho_n \sim 1\cdot3\times10^{14}$ g cm^{-3}. Surprisingly, in the entire range of densities where a lattice and transverse shear velocity can exist, the velocity c_s is almost constant although the density varies by a factor of 10^7. We shall assume it to have the constant value $c_s = 5\times10^7$ cm s^{-1} for all densities up to $\rho = 1\cdot3\times10^{14}$ g cm^{-3}. Then the characteristic frequencies of a torsionally vibrating non-rotating star depend only on its radius R. The lowest frequency torsional mode is a shape preserving twisting motion in which the northern and southern

degenerate sea of neutrons at temperature T would have viscosity η approximately given by[18]

$$\eta \sim \frac{p_f}{\sigma}\left(\frac{E_f}{k\,T}\right)^2 \tag{5}$$

where p_f is the neutron Fermi momentum, E_f the corresponding energy and σ the n–n scattering cross-section. Then for such a neutron fluid with $\rho \sim 10^{14}$ g cm^{-3} and $T \sim 5\times10^8$ °K, $n \sim 10^{14}$ and the torsional oscillation of the mantle would be damped in about 10^6 cycles if the oscillating mass were, say, one-tenth that of the whole star.

Such fluid core damping would be more than 10^6 times too fast if the pulsing radio sources are to be associated with free stellar torsional vibrations. Such rapid damping of the torsional vibrations of medium and heavy neutron stars is greatly reduced by two circumstances. The lattice is expected to disappear when nuclei will no longer condense out at a neutron density somewhat less than that of the neutrons within conventional nuclei. But in just this density regime the homogeneous nucleon fluid will behave as a superfluid. Both neutrons and protons at the top of their Fermi seas will be strongly attracted to like particles of opposite momenta and spin. (At higher neutron density and Fermi energy the neutron repulsive core begins to dominate the longer range attraction, at least for $^1S_0$ scattering.) The strong attraction between such fermions is the sufficient condition for the formation of a large BCS gap[19] exactly analogous to that in a superconductor, except that the neutrons and electron screened protons carry no net charge. For the calculated limiting density $\rho_n \sim 1.3 \times 10^{14}$ g cm^{-3} for the termination of nuclei condensation and lattice formation, the BCS energy gap for the neutrons has been calculated[20] to be $\sim 3 \times 10^6$ eV. This corresponds to a transition temperature into the superfluid state of 3×10^{10} °K, more than fifty times the expected temperature in a cooling neutron star. The proton component with a density a few hundred times less than that of the neutrons has an estimated energy gap[20] near 5×10^5 eV.

Although necessarily cavalier theoretical approximations may preclude quantitative accuracy in such estimates, it is very probable that the crystalline mantle of a neutron star rests on a nucleon superfluid much below its transition temperature[21,22]. The great reduction that ensues in the probability for single particle nucleon excitations above the gap causes a similar suppression of related dissipative effects in the superfluid. Equation (5), for example, is no longer relevant for the description of the neutron fluid viscosity. The transfer of momentum from the crystalline mantle to the adjacent superfluid can be a very complicated and slow process which involves the formation of quantized vortices in the superfluid and interactions with a small density of remnant normal fluid (really phonons) analogous to that in rotating containers of liquid HeII[23]. It seems at least plausible that the superfluid lubricant may very effectively suppress fictional losses between the core and the mantle. (The very degenerate electron component of the neutron–proton sea is not a superconductor but does have a very high electrical conductivity. It is restrained from carrying away momentum from the mantle because any net momentum in the electron component alone means a huge electric current which, even at frequencies $\omega \sim 10$ s^{-1}, cannot penetrate significantly into the highly conducting medium.)

A second factor which can suppress fictional damping from a liquid core is the large angular momentum which a neutron star would probably have, at least when it is formed. Rapid spinning can reduce the central pressure and, therefore, density to such an extent that the lattice structure could be maintained throughout the star even for heavier neutron stars. For a star rotating so fast that it begins to approach a more disk-like shape, the lowest torsional mode has a displacement proportional to $J_1(3.8\ r/R_D)$ in which the inner and outer parts rotate oppositely about the axis of rotation. The characteristic frequency $0.60\ c_s R_{\bar D}^{-1}$ can be very considerably less than that of the non-rotating sphere when the disk radius, R_D, is sufficiently greater than the sphere's.

Most of the volume of an oscillating shell is occupied by matter with $\rho \sim 10^{14}$ g cm^{-3}. In just this regime the calculation of c_s is least certain. The calculation of c_s for this density (Table 1) suggests that the characteristic torsional vibration periods based on an assumed $c_s = 5 \times 10^7$ cm/s may be a considerable underestimate.

I thank Professors M. Blackman, J. Dungey, M. Nauenberg, R. Orbach and Dr J. Park for discussions and Professor P. T. Matthews and the theoretical physics group at Imperial College for their kind hospitality.

This research was sponsored in part by the Air Force Office of Scientific Research, OAR, through the European Office of Aerospace Research, US Air Force.

Received May 30, 1968.

[1] Mestel, L., and Ruderman, M., *Mon. Not. Roy. Astro. Soc.*, **136**, 27 (1967).
[2] Tsuruta, S., and Cameron, A., *Canad. J. Phys.*, **44**, 1836 (1966).
[3] Bahcall, J. N., and Wolf, R. A., *Phys. Rev.*, **140**, B1452 (1965).
[4] Finzi, A., *Phys. Rev.*, **137**, B472 (1965).
[5] Meltzer, D., and Thorne, K., *Ap. J.*, **145**, 514 (1966).
[6] Hewish, A., Bell, S., Pilkington, J., Scott, P., and Collins, R., *Nature*, **217**, 709 (1968), (Paper 1).
[7] Wigner, E., *Phys. Rev.*, **46**, 1002 (1934).
[8] Fuchs, K., *Proc. Roy. Soc.*, **151**, 585 (1935).
[9] Clark, C., *Phys. Rev.*, **109**, 1133 (1958).
[10] Lindemann, F., *Phys. Z.*, **11**, 609 (1910).
[11] Pines, D., *Elementary Excitations in Solids*, Benjamin, New York (1963).
[12] Brush, S., Sahlin, H., and Teller, E., *J. Chem. Phys.*, **45**, 2102 (1966).
[13] Tsuruta, S., and Cameron, A., *Canad. J. Phys.*, **43**, 2056 (1965).
[14] Harrison, B., Thorne, K., Wakano, M., and Wheeler, J., *Gravitation Theory and Gravitational Collapse* (University of Chicago, 1965).
[15] Weiss, R., thesis, New York Univ. (1968).
[16] Wolf, R., *Ap. J.*, **145**, 834 (1966).
[17] Tsuruta, S., and Cameron, A., *Canad. J. Phys.*, **44**, 1895 (1966).
[18] Abrikosov, A., and Khalatnikov, I., *Reports on Progress in Physics XXII* (The Physical Society, London, 1959).
[19] Bardeen, J., Cooper, L., and Schrieffer, J., *Phys. Rev.*, **108**, 1175 (1957).
[20] Kennedy, R., Willets, L., and Henley, E., *Phys. Rev.*, **133**, B1131 (1964).
[21] Ginzburg, V., and Kirzhnita, *Sov. Phys. JETP*, **20**, 1346 (1965).
[22] Ruderman, M., *Fifth Eastern Theoretical Physics Conference Proceedings*, edit. by Feldman, D. (Benjamin, New York, 1967).
[23] Androninkashvili, E., and Mamaladze, Yu, *Progress in Low Temperature Physics*, **5** (North-Holland, Amsterdam, 1967).

36. Possible Interpretation of Pulses from a Radio Source

by
M. R. KUNDU
R. M. CHITRE
Tata Institute of Fundamental Research,
Bombay, India

THE Cambridge group[1] recently announced the recording of pulses from a local object, lasting for about 0.016 s and repeating with extreme regularity with a period of 1.337 s, the accuracy being one part in 10[7]. There have been further observations from Cambridge[2] and from Jodrell Bank[3]. The amplitude of the pulses varies randomly, the pulses appearing and disappearing for periods of a few minutes. There is also a fine structure superimposed on the main pulse. Despite the regularity of the pulses, the power emitted varies significantly over all periods. Furthermore, observations indicate a frequency drift of -5 MHz s^{-1}. The absence of any proper motion of the source, and the interpretation of the frequency drift in terms of dispersion through the interstellar plasma, limit the distance of the source to the range 10^3 A.U. $< d < 65$ pc. An eighteenth magnitude blue stellar object has been found near the object, although there seems to be uncertainty about the nature of the object. From the pulse width and the rate of frequency drift the source is smaller than 5×10^8 cm.

Hewish *et al.*[1] tentatively suggested that the radio signals come from a pulsating neutron star or a white dwarf, and Ryle and Bailey[2] suggested the possibility of coherent plasma oscillations. Saslaw *et al.*[4] argue that the periodicity of the pulses could arise from the rotation of a binary system of neutron stars, the gravitational lens effect focusing the radio waves during the rapid rotation. We are not concerned with the regularity of the pulses; they probably originate in the vibrations of a superdense object such as a neutron star or a white dwarf which is capable of storing energy of $\sim 10^{51}$ ergs in the vibrational mode. We shall rather assume the existence of the pulse regularity and attempt to account for the impulsive radiation in terms of plasma oscillations originating in the magnetosphere of a neutron star. The existence of such a magnetosphere around a neutron star has been discussed by Cameron[5]. The fine structure that is often superimposed on the main pulse is evidently reminiscent of repeated excitations of plasma waves. Following the suggestion of Cameron, large magnetic fields ($\sim 10^{10}$ gauss) prevailing on these stars may be twisted by the rotation to form arches. It is conceivable that there are neutral zones in the field, just as on the Sun where the field pinches give rise to a flare type of activity. During each vibration of the star, hydromagnetic shock waves will be set up, and particles released as a result of the pinch will be accelerated. These particle streams or the shock waves themselves may excite the plasma oscillations. The number density of trapped electrons could be in the range 10^6–10^8 cm^{-3}, corresponding to plasma frequencies of 10–100 MHz. Such electrons may come from the abundant concentration of neutrons and the relatively very much lower density of protons.

It is obvious that synchrotron radiation cannot account for these pulses, because in a high magnetic field such as exists in a neutron star the radiation would be damped out in a very small fraction of a second. The synchrotron mechanism has also been ruled out by Ryle and Bailey[2] and by Lyne and Smith[6] because of the very high surface brightness and the absence of a low-frequency cut-off. Because the magnetic field is high, one ought to consider gyro harmonic-resonance absorption of the radio waves. Our calculations indicate that with a dipole type of field, with $|H| \propto 1/r^3$ (r measured from the centre of the star), the gyro harmonic-resonance levels would not occur until a distance of approximately 300–400 stellar radii is reached, where the electron density would be extremely small and therefore the absorption would not be significant. Besides, at such a distance one does not know what happens to the field of the star; it is probably tangled with the interstellar field and it does not seem meaningful to consider gyro harmonic-resonance absorption at that level.

The observed frequency drift from high to low frequencies has been attributed by Hewish *et al.*[1] as well as by the Jodrell Bank group to dispersion through the interstellar plasma, although there is always the possibility that part of the drift could originate in the source itself. Indeed, Pilkington *et al.*[7] report that the frequency drift rate ranges from 5 MHz s^{-1} to 20 MHz s^{-1}. It seems to be difficult to explain such a large variation of drift rate on the basis of dispersion through the interstellar plasma; part of the dispersion probably arises in the source itself. The frequency drift from the source may be the result of an exciting disturbance passing through different plasma levels in the star's atmosphere. The scale height in the atmosphere of such a massive object is $kT/M_n g$ (M_n being the neutron mass and g the gravitational acceleration) and for a star with $M \simeq M_\odot$, $R \simeq 5 \times 10^8$ cm, it comes out as $\sim 10^6$ cm.

If we consider, for example, two plasma levels corresponding to 100 and 105 MHz, the levels are separated by about 10^7 cm. Because the typical Alfvén speed in the atmosphere is about 10^7 cm s^{-1}, the time of travel between the two levels is approximately 1 s, which will probably account for the observed drift. If, instead of the shock waves, the particle streams excite the plasma waves, then the corresponding time of travel becomes smaller by an order of magnitude, because the velocity of the stream has to be at least 10^9 cm s^{-1}. The fine structure superimposed on the main pulse could be caused by repeated pinches which might occur each time the plasma oscillations are excited. It is tempting to suggest that the pulses are similar to solar bursts of type III or type I. It is reasonable to expect that the magnetic field of the star will be twisted into loops because of the rapid rotation and pulsation. One might therefore expect pinches to occur in one or other of the looped fields in localized regions each time the star vibrates. This picture is consistent with the recent interpretation by Drake that the fine structure radiation originates in a source of a few hundred km in size. As Hewish *et al.*[1] pointed out, the energy requirements for such a model are adequately satisfied because of the high energies involved in the vibrational modes.

The linear polarization observed by the Jodrell Bank group[6] may be caused by the existence of a polarization limiting region in the star's outer magnetosphere. The polarization limiting region occurs[8] where the coefficient of coupling between the two magneto-ionic modes, given by $Q = 10^{-17} f^4 N^{-1} H^{-3} S^{-1}$ (f is the frequency, N the electron density, H the magnetic field, and S the scale height of the magnetic field), is unity. A single magneto-ionic mode proceeding from a region of weak coupling to one of strong coupling retains the polarization appropriate to the point $Q = 1$. We may assume that circular polarization is initially imposed on the radio waves as a result of selective absorption of one of the magneto-ionic modes, namely, the extraordinary mode. The absorption can conceivably be very large for the extraordinary mode in the presence of a high magnetic field such as exists in the atmosphere of a neutron star. The polarization changes from circular to linear as the radiation passes through a quasi-transverse (QT) region of the magneto-ionic medium. The condition for the linearity to occur is given by $Q = 1$ in the QT region. It is evident that near the surface Q is very much smaller than 1 because of the very high field, whereas the region where $Q = 1$ lies quite far from the surface, at a distance of approximtely 100 stellar radii. This is because the electron density must decrease, exponentially say, away from the surface. Suppose Q becomes equal to 1 at a distance where the electron density is (say) 10^2–10^3 cm^{-3}, then the corresponding value of H is about 100–200 gauss, which occurs at around 100 stellar radii assuming a dipole field approximation for fields far from the star. This distance is large compared with the size of the star and therefore the magnetic field lines which are likely to be twisted near the surface are not likely to be affected significantly at these large distances. As a result, the radiation originating in one or other of the small regions or pinches around the star will always propagate nearly at right angles to the field in the QT region (propagation at right angles is necessary because the QT approximation holds only for a very small range of angles near 90°), giving rise to the linear polarization of the pulses. Beyond this region, $Q > 1$, so the character of the polarization does not change.

We thank Dr S. Swarup for many helpful discussions.

Received May 2, 1968.

[1] Hewish, A., Bell. S. J., Pilkington, D. H., Scott, P. F., and Collins, R. A., *Nature*, 217, 709 (1968), (Paper 1).
[2] Ryle, M., and Bailey, J. A., *Nature*, 217, 907 (1968), (Paper 19).
[3] Davies, J. G., Horton, P. W., Lyne, A. G., Rickett, B. J., and Smith, F. G., *Nature*, 217, 910 (1968), (Paper 6).
[4] Saslaw, W. C., Faulkner, J., and Strittmatter, P. A., *Nature*, 217, 1222 (1968), (Paper 42).
[5] Cameron, A. G. W., *Nature*, 205, 787 (1965).
[6] Lyne, A. G., and Smith, F. G., *Nature*, 218, 124 (1968), (Paper 7).
[7] Pilkington, J. D. H., Hewish, A., Bell, S. J., and Cole, T. W., *Nature*, 218, 126 (1968), (Paper 2).
[8] Cohen, M. H., *Ap. J.*, 131, 664 (1960)

37. Radio Pulse Profiles from Pulsating White Dwarfs

by

B. H. BLAND

Department of Computer Science,
University of Manchester

FOLLOWING the publication of well resolved mean pulse profiles[1] for the four pulsating radio sources discovered by Hewish *et al.*[2,3], an attempt has been made to reproduce the profiles from simple models of pulsating white dwarfs. The calculated profiles are developed from the observation that the pulse length is comparable with the light time of the radius of a typical white dwarf, but include also the effects of a spatial and temporal distribution of emission over the surface.

The excitation of the emission is assumed to be an impulse, which may, for example, be a pressure wave arising from the pulsation steepening into a shock as it reaches the surface. Radial harmonic modes are expected to be excited in preference to the fundamental because the pulsation is being driven from near the surface, and because dissipation in the degenerate core is less[4,5]. The excitation does not, however, reach the whole surface simultaneously, for the star may be rotating sufficiently to give an appreciable gradient of surface gravity from pole to equator. Because the natural period depends on the surface gravity[4] there will be a phase lag in the pulsation according to latitude. At colatitude θ the pulse is assumed to reach the surface at a time $t_e = \tau \sin^2\theta$ after that for the poles, where τ is the time taken for the wave of emission to travel from pole to equator. A phase lag of some kind is necessary to explain why the short period source $CP0950$ does not have a shorter pulse length than the other pulsars.

In the model the shape of the pulse is determined by: (i) the time (t_e) at which each surface element radiates; (ii) the relative strength (f_e) of the emission from each element; (iii) the shape of the emitting surface, which determines the light time to the observer; (iv) the directional dependence (f_a) of radiation from each element.

The light time is calculated as for a sphere viewed at an angle α to the equatorial plane, so that radiation from a surface element at colatitude θ and longitude φ arrives at a time

$$t = t_0 + t_e - \frac{R}{c}(\sin\alpha\,\cos\theta + \cos\alpha\,\sin\theta\,\cos\varphi)$$

where t_0 defines the leading edge of the pulse profile and R is the radius of the star. The flux received from this element is

$$dF = f_a\,f_e\,R^2\,\sin\theta\,d\theta\,d\varphi$$

The exact forms of f_a and f_e appropriate to a single pulse may be quite complicated and variable, but their time averages are taken to be simple functions of α, θ, and φ only, which may represent, for example, the average way in which the magnetic field strength and direction vary with latitude.

Models of the Emission

The objective was to search for physically reasonable forms of f_a and f_e which reproduced all the observed pulse profiles, allowing only a variation in the delay τ, the inclination α and the radius R. The distribution of inclinations derived from the known sources should also be consistent with a random orientation of sources in space. In the integration to obtain the pulse shape, time was expressed in units of R/c, with a parameter $K = \tau c/R$ representing the delay. If a match can be achieved between the computed and observed pulse shapes, then

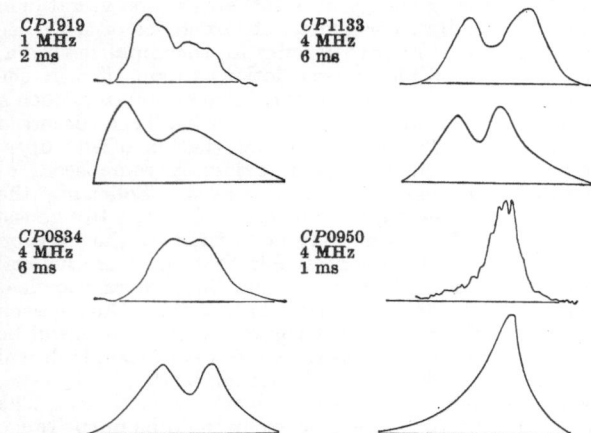

Fig. 1. Observed and computed pulse profiles for the four Cambridge sources. The observed profile is the upper curve of each pair.

the radius of the source is c times the observed pulse length divided by the computed dimensionless pulse length. The mean pulse profile is observed to be constant in time[1] so that $f_e f_a$ must be axisymmetric, which rules out oblique rotator models. It was assumed that f_e is likely to be proportional to the mean square of some component of the surface field. The forms of f_e which were first tried corresponded to a dipole magnetic field along the polar axis: (i) $f_e = \sin^2\theta$ for the horizontal component; (ii) $f_e = \cos^2\theta$ for the vertical component; (iii) $f_e = 1 + 3\cos^2\theta$ for the total field; (iv) a uniform distribution $f_e = 1$ was also tried.

Fifteen forms of the angular function f_a were tried: $\cos\delta$, $\cos^2\delta$, $\cos^3\delta$, 1, $\sin\delta$, $\sin^2\delta$, $\sin^2\delta\cos^2\delta$, $\cos^2\omega$, $\cos^4\omega$, $\cos^2\omega\cos\delta$, $\cos^2\omega\cos^2\delta$, $\sin^2\omega$, $\cos^2 q$, $\sin^2 q$, $\cos^2\delta\,(\cos^2 q\sin^2\delta + \cos^2\delta)$, where δ is the angle between the line of sight and the normal to the surface, ω is the angle between the line of sight and the dipole field line, and q is the azimuth of the line of sight with respect to the field line.

The four emission functions f_e were permuted in pairs with the fifteen angular functions f_a, and for each pair pulse profiles were computed for eight values of α and seven values of K. In all, about 3,000 profiles were produced. Despite the variety of functions chosen, none of the sixty pairs yielded pulse shapes resembling the observed profiles. The profiles obtained suggested that the failure to match the observations was due to f_e rather than f_a.

Setting the angular dependence to the simple form $f_a = \cos\delta$ corresponding to zero limb darkening, the form of emission variation $f_e = \sin^4\theta\,\cos\theta$ was found to yield

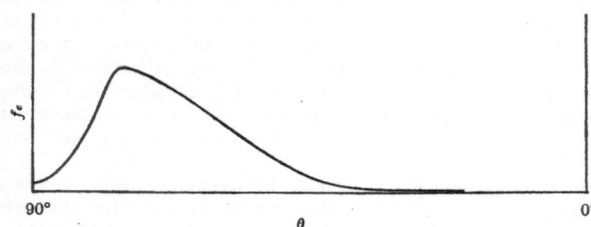

Fig. 2. The relative distribution of radio emissivity (f_e) as a function of the polar angle θ. Symmetry about the equatorial plane is assumed.

Table 1

Source	p (s)	Pulse length (ms)	R (km)	K	a	$M/M_\odot$	F/F_0
CP1919	1·33	47	14,300 ± 1,200	0·18 ± 0·1	38 ± 3	0·21 ± 0·05	0·85
CP0834	1·27	59	13,300 ± 1,400	0·76 ± 0·1	40 ± 3	0·23 ± 0·05	0·83
CP1133	1·18	61	14,600 ± 1,400	0·67 ± 0·1	30 ± 4	0·20 ± 0·05	0·90
CP0950	0·25	58	6,900 ± 1,500	3·70 ± 1·1	18 ± 10	0·85 ± 0·17	0·95

reasonable profiles. Improvements were then made by an iterative process in which f_e was defined as a numerical function of θ. The success of this matching process is shown in Fig. 1, which compares the computed profiles with the experimental profiles published by Lyne and Rickett[1]. Values of R, K and α for the four sources are given in Table 1. The chosen function $f_e(\theta)$ is shown in Fig. 2. The computed profiles show all parts of the pulse for which the intensity is above 2 per cent of the maximum.

Remaining discrepancies which appear to be significant are that in the model the separation of the double peak is too small for CP1133 and too great for CP0834. The separation of the double peaks arises from the light travel time between the two active regions, so that a small adjustment of the latitude of these regions would improve the fit. A peak emission at latitude $\beta_m = 24°$ for CP1133 and at $\beta_m = 11°$ for CP0834 provides a good match, as compared with a maximum at 18° for the model used for Fig. 1.

Interpulse from CP0950

The weak pulse observed by Rickett and Lyne[6], occurring 100 ms before the main pulse of CP0950, provides striking confirmation of the model, for the large value of phase delay ($K = 3·7 \pm 1·1$) shows that excitation first reaches the polar region 80 ± 20 ms ahead of the main peak of the profile. Radiation from the polar region is emitted in a short pulse since the velocity of the wave of emission, $c/K \sin\theta \cos\theta$, is high when θ is small. The value of f_e near the poles must be increased slightly to account for the strength of the prepulse, which is 1·8 per cent of

that of the main pulse. An improved fit for the whole profile of CP0950 is now obtained by setting the maximum of emission at latitude 14°, with $K = 4·7$ and $α = 30°$.

The other three sources have smaller values of K, and no interpulse of this kind would be seen clearly distinguished from the main pulse. The interpulse would be advanced on the main peak by 5 ms for CP1919, 20 ms for CP1133, and 40 ms for CP0834, where it forms part of the rising front of the pulse.

Magnetic Field Structure

It is remarkable that the variation of emissivity f_e with latitude is very similar to the distribution of magnetic field on the Sun[7]. This is not to suggest that the radio emission mechanisms are related, but rather that the underlying magnetohydrodynamics responsible for the field structure is the same. The subpulses which build up the pulse profile may be a result of radiation from local regions of higher magnetic field, like those associated with sunspot groups. It is tempting to speculate that magnetic stars may also have this same field structure.

The interpulse from CP0950 is very highly linearly polarized, and because this originates from the pole of the star it suggests that there may be a weak dipole field in addition to the field concentrated in the belt.

The range of latitude at which the maximum occurs, from 11° to 24°, suggests that there might be a variation resembling that of the sunspot cycle. If this occurs, it should become noticeable as a change of pulse shape or as a change in the radio polarization. Optically there may be changes in brightness and spectrum, as in magnetic stars.

Source Size

The most valuable parameter that can be derived from the analysis of the pulse is the star's radius. The radius is proportional to the length of the pulse and it is first necessary to remove the broadening effects of the variable

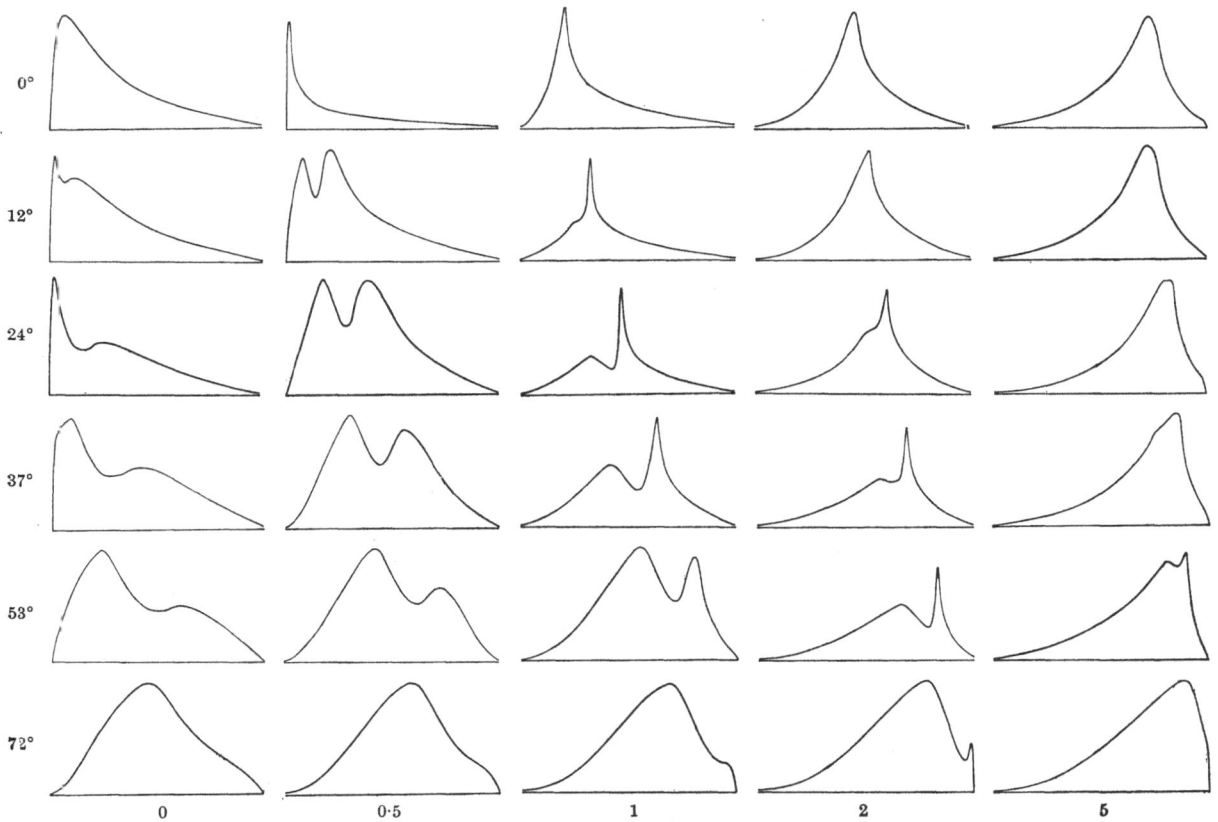

Fig. 3. Illustration of the variation of pulse shape with increasing rotation, $K = 0$ to 5, and increasing inclination to the line of sight, $a = 0°$ to 72°. although the pulse length and flux are functions of a and K, each profile has been scaled to the same length and width to facilitate comparison.

dispersion across the receiver bandwidth, the finite sampling interval of between 1 ms and 6 ms, and the error in estimating the pulse repetition period during the observation. Pulse lengths at the 2 per cent intensity level are given in Table 1. The profiles measured by Lyne and Rickett[1] are averages over a large number of polarization angles. This averaging is essential because different features within the pulse, corresponding to different areas on the source, have different polarizations. For the purpose of comparing pulse shapes the observed profiles are accurate to between 5 per cent and 10 per cent.

The standard errors accompanying the radii in Table 1 represent the uncertainty involved in adjusting α and K to match the observed pulse shapes, together with the error in the pulse length. Errors arising from imperfections in f_a and f_e are probably of the same order. Solutions with $f_a = \cos^2\delta$ yield very similar results.

When $K \gg 1$ the pulse shape is dominated by the phase lag τ, so that the solution for $CP0950$ is not as strong as for the more slowly rotating sources. The evolution of the pulse shape with increasing rotation and inclination is shown in Fig. 3. Inclinations of $0°$ to $53°$ have been chosen at intervals of equal probability of observation, while $\alpha = 72°$ illustrates the appearance at high inclination, the probability of finding a source with $\alpha > 72°$ being less than 5 per cent. The apparent radio luminosity of the source is a function of the inclination and the final column of Table 1 contains the flux received relative to that at $\alpha = 0$.

The fact that the derived radii fall comfortably within the range of known white dwarf radii[8] is independent evidence in support of the model. The radius of a white dwarf is not very sensitive to temperature or composition so that a good estimate of the mass can be obtained once the radius is known. The masses M in Table 1 and the central densities in Table 2 are from the series of helium models published by Skilling[9]. The mean density $\bar{\rho}$ and the surface gravity g follow from the mass and radius. It is perhaps worth noting that $p \bar{\rho}^{\frac{1}{2}}$ is essentially constant. The overtone number n has been calculated from Faulkner and Gribbin's models[10]. It is encouraging to find that, within the errors, all four sources are pulsating in the same harmonic. For comparison, there is some evidence of $n = 14$ harmonic oscillations in an old nova[11]. If the surface layers of the star are to keep in step with the pulsation, then the peak to peak amplitude must not exceed the distance that material can fall in half the period, that is $\leqslant 1/8\, g\, p^2$, which sets a limit of the order of 30 km for all four sources. Consequently, any optical variability is likely to consist of flashes generated by the shocks, with a negligible sinusoidal component.

Rotation

The second parameter in the solution, K, is directly related to the star's angular velocity of rotation ω. Provided that the time lag is small compared with the period p it is reasonable to expect τ/p to be proportional to the ratio of centrifugal acceleration to surface gravity:

$$\frac{\tau}{p} = \frac{Q\, R^3\, \omega^2}{GM}$$

where Q is some dimensionless constant that depends on the degree of coupling between different parts of the star, and perhaps also on the harmonic n. The strength of coupling can only be derived from a detailed physical study of the pulsation cycle. The dissipative part of the coupling is know to be small[5] and Q is likely to be large. To show the order of magnitude involved one may put

$Q = 100$ which leads to a rotation period of 2 h for $CP1919$ and 4 min for $CP0950$.

If the coupling factor is the same for two sources, then the ratio of their angular velocities is

$$\left(\frac{\omega_1}{\omega_2}\right)^2 = \frac{p_2 K_1 M_1}{p_1 K_2 M_2}\left(\frac{R_2}{R_1}\right)^2$$

Angular velocities relative to $CP\,1919$ are given in Table 2. The angular velocity increases as the radius decreases, as one would expect, although four sources is a small sample. The condition $\tau \ll p$ is beginning to break down for $CP0950$ and the phase lag may no longer be linearly related to ω^2. If $CP0950$ is not to exceed the break-up limit, its rotation period must be longer than 40 s, so that the rotation period of $CP1919$ is unlikely to be shorter than 30 min.

The arrival time of an individual subpulse is $(R/c)\cos\alpha \sin\theta \cos\omega t$, and if such pulses arise from discrete quasi-stationary regions then the motion of these subpulses within the main pulse as the star rotates could be used to deduce the rotation period independently of any theory about the pulse shape. The most promising method of determining the period of rotation is from an auto-correlation of the radio flux. That nothing has been detected so far is probably a result of the strong super-imposed scintillation[12-14].

Distance of the Sources

If a source can be identified with a visible star, then the surface temperature can be combined with the apparent magnitude and the radius, as determined in this paper, to calculate the distance. We may expect the pulsars to have similar temperatures, especially the three which have closely similar radii. If $CP0950$ also has a similar surface temperature, its luminosity will be 1·5 magnitudes fainter, but because the measures of dispersion in the ionized interstellar medium suggest a much smaller distance then we can expect all four pulsars to be identifiable at least as easily as $CP1919$. The fields of the three pulsars $CP0950$, $CP0834$, $CP1133$ contain nothing but faint objects (personal communication from W. Sargent), so that the identification of $CP1919$ by Ryle and Bailey[15] must be left in doubt.

If $CP0950$ is at a distance of the order of 20 pc (ref. 3), then its surface temperature must be very low, which provides a restriction on the type of energy source responsible for maintaining the pulsation. If $CP1919$ and $CP0834$ lie at distances of the order of 100 pc, then it may be possible to observe pulsars in the nearest globular clusters. This would provide a calibration of the absolute radio flux as a function of period, a test of the uniformity as well as the actual value of the interstellar electron density, and some information as to their evolutionary history.

I thank Professor F. D. Kahn, Professor F. G. Smith, and Mr E. Graham for stimulating discussions, and A. Lyne and B. Rickett for making their observations available before publication.

Received June 24, 1968.

[1] Lyne, A. G., and Rickett, B. J., *Nature*, **218**, 326 (1968), (Paper 10).
[2] Hewish, A., Bell, S. J., Pilkington, J. D. H., Scott, P. F., and Collins, R. A., *Nature*, **217**, 709 (1968), (Paper 1).
[3] Pilkington, J. D. H., Hewish, A., Bell, S. J., and Cole, T. W., *Nature*, **218**, 126 (1968), (Paper 2).
[4] Kahn, F. D., and James, R. A., *Meeting of the Royal Astro. Soc.* (London, April 10, 1968).
[5] Thorne, K. S., and Ipser, J. R., *Ap. J. Lett.*, **152**, L71 (1968).
[6] Rickett, B. J., and Lyne, A. G., *Nature*, **218**, 934 (1968), (Paper 11).
[7] Ringnes, T. S., *Astrophys. Norv.*, **10**, 189 (1968).
[8] Eggen, O. J., and Greenstein, J. L., *Ap. J.*, **141**, 83 (1965).
[9] Skilling, J., *Nature*, **218**, 923 (1968), (Paper 31).
[10] Faulkner, J., and Gribbin, R., *Nature*, **218**, 734 (1968), (Paper 28).
[11] Lawrence, G. M., Ostriker, J. P., and Hesser, J. E., *Ap. J. Lett.*, **148**, L162 (1967).
[12] Smith, F. G., *Nature*, **218**, 720 (1968), (Paper 24).
[13] Scheuer, P. A. G., *Nature*, **218**, 920 (1968), (Paper 46).
[14] Tanenbaum, B. S., Zeissig, G. A., and Drake, F. D., *Science*, **160**, 760 (1968).
[15] Ryle, M., and Bailey, J. A., *Nature*, **217**, 907 (1968), (Paper 19).

Table 2

	ρc g cm⁻³	$\bar{\rho}$ g cm⁻³	$p\bar{\rho}^{\frac{1}{2}}$	g cm s⁻²	τ ms	ω/ω $CP1919$	Harmonic n
$CP1919$	$1\cdot7 \times 10^5$	$3\cdot4 \times 10^4$	250	$1\cdot4 \times 10^7$	9	1	11
$CP0834$	$2\cdot7 \times 10^5$	$4\cdot6 \times 10^4$	270	$1\cdot6 \times 10^7$	34	$2\cdot4$	10
$CP1133$	$1\cdot5 \times 10^5$	$3\cdot0 \times 10^4$	210	$1\cdot3 \times 10^7$	32	$1\cdot9$	13
$CP0950$	$1\cdot1 \times 10^7$	$1\cdot2 \times 10^6$	270	$2\cdot4 \times 10^8$	85	44	13

38. Pulsar Models

by

W. J. COCKE
JEFFREY M. COHEN
Institute for Space Studies,
Goddard Space Flight Center,
New York

IN this article we examine the possibility, previously considered by others, that the pulsed radio sources observed recently are pulsating white dwarfs. We first present the results of radial mode eigenvalue calculations (for various central densities and compositions) which were obtained by integrating the eigenvalue equations themselves rather than by using variational techniques. Computations of non-linear white dwarf pulsations are then presented, in which a hydrodynamic code was used to follow pulsations of finite amplitude driven by pycnonuclear reactions. Very large amplitude pulsation of neutron stars is also considered. The possibility that the optical pulsation periods are twice as long as the radio periods has arisen from observations[1]. We discuss a method for such a "frequency doubling" based partly on magnetohydrodynamics.

Eigenvalue Calculations

The following white dwarf eigenfrequency calculations were obtained by computer integration of the Newtonian equations for the eigenfunctions. The eigenvalues were obtained by finding those values of the period for which the eigenfunction $\zeta = \delta r/r$ and $d\zeta/dr$ are finite at the outer boundary and match to an analytic expansion (about the surface) of the eigenfunction. The eigenvalues compare very well with those determined by variational methods (S. Vila at the New York Conference on Pulsars, and refs. 2 and 3), but the method described here has the advantage of giving accurate eigenfunctions without having to postulate polynomial forms. In Table 1 we have used the degenerate electron equation of state as first applied to white dwarfs by Chandrasekhar[4]. The effect of Hamada–

Salpeter corrections to the equation of state[5] and of other corrections reduces the periods slightly. In Table 2 we show the results of calculations corrected for the Coulomb, Thomas–Fermi and exchange interactions (Hamada–Salpeter corrections). In both tables the ratio of nucleons to electrons was taken to be 2·0. Thus the uncorrected models can represent, for example, C^{12} or $C^{12} + He^4$ compositions while the corrected model assumes $Z = 6$ as with C^{12}.

For higher densities than those considered here, relativistic effects are too large to neglect (J. M. Cohen at the New York Conference on Pulsars, and refs. 2 and 6), and periods considerably lower than 1·64 s are very difficult to obtain. This difficulty, however, can be circumvented. We later give a mechanism which generates two radio frequency pulses for each stellar pulsation. Also in the following sections we discuss dynamical calculations with simulated nuclear excitations, which tend to show that pulsation periods shorter than the fundamental may in any case be obtained. Thus there are at least two possible ways of producing models with radio periods shorter than the fundamental periods.

Dynamical Calculations

The full, non-linear hydrodynamical equations of motion were used to follow the oscillations of a white dwarf starting from initial conditions close to equilibrium. We use the Lagrangian form[7]

$$\frac{\partial^2 r}{\partial t^2} = -4\pi r^2 \partial_m (P+Q) - \frac{GM}{r^2}$$

where m is the mass inside the radius $r = r(m,t)$, p is the pressure, and Q is a viscosity term introduced both to spread out shock fronts over several mass zones, and to eliminate the effects of the initial conditions on the final state of motion of the white dwarf. We used 20 mass zones of equal mass, and used a form for Q given by Colgate and White[8] and Arnett[9] as $Q_{k+\frac{1}{2}} = C\rho_{k+\frac{1}{2}}(v_{k+\frac{1}{2}} - v_k)^2$ for $v_{k+1} < v_k$ and $Q_{k+\frac{1}{2}} = 0$ otherwise, where v_k is the velocity of kth mass skell, $\rho_{k+\frac{1}{2}}$ is the density, and C is a constant. Arnett[9] chooses $C = 1$, but we chose $C = 10$ in order to damp out the effects of the initial conditions in a reasonable computer time.

The pressure is given by $P = P_e + P_n$, where P_e is the usual degenerate electron pressure and P_n is due to pycnonuclear reactions[10]. We have used the form $P_n \propto (\rho - \rho_1)^2$, where ρ_1 is a density at which the excess pressure from the heating from the nuclear reactions is assumed to die away. The reactions were assumed to begin at a density $\rho_2 > \rho_1$. The total mass in the model used was assumed to be $M = 8·38 \times 10^{32}$ g, corresponding to a central density ρ_c of about 2×10^6 g/cm³. Of course, no pycnonuclear reactions actually occur at this density. We would have liked to have run models with densities high enough for these reactions to occur, but we found that for $\rho_c \gtrsim 10^7$ g cm⁻³, the initial deviations from equilibrium of our model gave rise to gravitational collapse! One can imagine that the general results obtained would be applicable to more compact configurations which

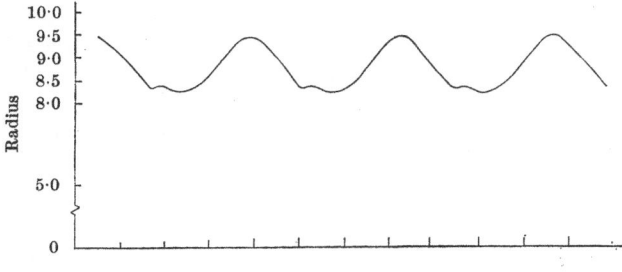

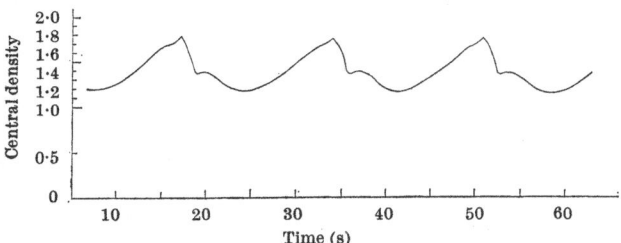

Fig. 1. Outer radius (in units of 100 km) and central density of a white dwarf (in units of 10^6 g cm⁻³) versus time.

will not collapse for sufficiently small oscillations. We conclude, however, that large amplitude oscillations would result in collapse for the more compact configurations, at least down to the neutron star stage, as we will discuss.

We have assumed that the nuclear energy source is very close to the surface where, for example, He^4 might be converting to C^{12}. Using the form $P_n = C_0(\rho - \rho_1)^2$ for the over-pressure in the penultimate mass zone, we found that $C_0 = 9.0 \times 10^{11}$ and $\rho_1 = 5.0 \times 10^4$ give reasonable oscillations with $\rho_2 = 7.0 \times 10^{14}$. Fig. 1 shows the final asymptotic time dependence of the outside radius and the central density. The time intervals between the large radius maxima are nearly equal to the fundamental frequency for small pulsations. Small secondary maxima also appear. In more realistic models, the secondary maxima might well be larger, thus shortening the apparent oscillation period. Many optical variable stars have been noted to have periods shorter than the fundamental[11].

Similar calculations were also carried out for very large amplitude pulsations of degenerate neutron stars. For a neutron star of total mass $M = 1.0 \times 10^{33}$ g, an oscillation period of 1.33 s can be obtained if the maximum radius is taken to be 2,600 km. In Fig. 2 we have plotted the outside radius as a function of time; the period agrees with the estimates of Hoyle and Narlikar[12]. Even if their mechanism to avoid high energy loss by neutrinos at maximum contraction is possible, however, we find that the density at maximum extension is so low that free neutron decay takes place again resulting in neutrino loss. In fact, the damping time for the pulsations is of the order of a few months.

Radio Emission

In this section we discuss plasma mechanisms for generating the observed radio emissions. We assume that the basic energizing mechanism is an electric field in a layer where the scale height is smaller than the mean free path of thermal electrons. Thus we have a "low density" plasma in which ordinary magnetohydrodynamics is invalid, and the electric field may have a component parallel to the magnetic field[13]. From Faraday's law, $-\dot{B} = c\nabla \times E$, we see that maximum E is obtained for maximum $|\dot{B}|$, which corresponds to maximum $|d\rho/dt|$ in the deeper layers of the star, where magnetohydrodynamics is valid. We will see that fields $B \gtrsim 60$ gauss are required, and thus the magnetic pressure dominates the plasma pressure in the low density layer. Changes in B in the interior will then be communicated to the atmosphere. But max $|d\rho/dt|$ occurs twice during a pulsation cycle, and thus maximum E occurs twice during a cycle, making the stellar pulsation period twice the radio period.

The appearance of an electric field E with a component parallel to the magnetic field will then give rise to longi-

tudinal plasma oscillations in which the electrons move parallel to B, at the plasma frequency ω_p. These oscillations themselves do not radiate energy because there is no magnetic field associated with them, but in an inhomogeneous medium there is coupling with other radiative plasma wave modes[14], and the emitted radiation would then be at the plasma frequency ω_p. Kundu and Chitre[15] have also mentioned the possibility of plasma oscillations as an exciting mechanism.

Let us discuss radiation at 100 MHz. The corresponding plasma frequency is $2\pi \times 10^8$ s^{-1}, and the electron density is then $n = 1.24 \times 10^8$ cm^{-3}. Let us assume that the thermal temperature is 10^4 °K, and that the local density scale height is $l_0 = 6 \times 10^3$ cm, a typical scale height for a white dwarf atmosphere. The electron mean free path λ_e is then[16], with $\ln \Lambda = 15$ (where $\ln \Lambda$ is the Coulomb logarithm)

$$\lambda_e - 1.3 \times 10^5 T^2/n \quad \ln \Lambda = 7.2 \times 10^3 \text{ cm}$$

Thus, $\lambda_e > l_0$, and magnetohydrodynamics breaks down. The l_0 assumed here may actually be too small, because the intense radio output from the pulsar will probably distend the atmosphere. This same effect, however, will also raise the temperature, and we will still have $\lambda_e > l_0$.

We assume that the average power emitted from this layer is about 10^{28} ergs s^{-1} (ref. 17). The conversion from longitudinal oscillations to radiative plasma waves could occur in many (say 10^7) bursts per fundamental pulsation, so that if W is the energy stored in the plasma oscillations before release, $W \sim 10^{21}$ ergs. The number N of participating electrons is $N = 4\pi R_0^2 l_0 n$, and if $R_0 = 2.5 \times 10^8$ cm, $N \simeq 6 \times 10^{29}$. Thus each electron must possess, on the average, an energy w of 1.6×10^{-9} ergs. But if δr_{cs} is the charge separation associated with the oscillations and E is the electric field strength, we have from $E = en\delta r_{cs}$, $w = eE\delta r_{cs} = E^2/n$. Therefore $E = 0.42$ e.s.u.

We may now use Faraday's law to find the rate of change with time of the magnetic field by assuming various values for the scale δr_E of the electric field. One possibility is to set $\delta r_E \simeq l_0$, but for a large scale magnetic field, possibly like a dipole field, we might also set $\delta r_E \simeq R_0$. In any case, $\dot{B} \simeq cE/\delta r_E$, and for $\delta r_E = 6 \times 10^3$ cm, $\dot{B} \simeq 2 \times 10^6$ gauss s^{-1}, which implies a rather large field if $\dot{B} \simeq B$ for a radio pulsar oscillation period of about 1 s. For $\delta r_E = 2 \times 10^8$, $\dot{B} \simeq 60$ gauss s^{-1}, which is more reasonable. Note that if our estimate of l_0 is too low, then smaller values of $\dot{B}$ will suffice to produce the emission.

It is observed that the pulses have characteristically short rise times. We note that mode coupling from the plasma oscillations to radiative plasma waves is nonlinear and therefore short groups of bursts with rather sharply defined frequency spectra might be expected to occur.

Table 1. WHITE DWARF MODELS WITH DEGENERATE ELECTRON PRESSURE ONLY
(c.g.s. units)

Log density (g × cm⁻³)	Mass/10³³	Radius/10⁸	Period
10	2·81	1·31	1·64
9·5	2·78	1·81	2·07
9	2·68	2·46	2·63
8·5	2·54	3·28	3·40
8	2·32	4·30	4·49
7·5	2·00	5·55	6·11
7	1·60	7·04	8·67
6·5	1·17	8·81	13·0
6	0·785	10·9	20·5

Table 2. WHITE DWARF MODELS WITH HAMADA-SALPETER CORRECTIONS
(c.g.s. units)

Log density (g cm⁻³)	Mass/10³³	Radius/10⁸	Period
10	2·75	1·30	1·64
9·5	2·71	1·79	2·06
9	2·63	2·43	2·61
8·5	2·48	3·24	3·37
8	2·26	4·24	4·44
7·5	1·94	5·45	6·03
7	1·54	6·89	8·52
6·5	1·12	8·57	12·7
6	0·743	10·5	19·9

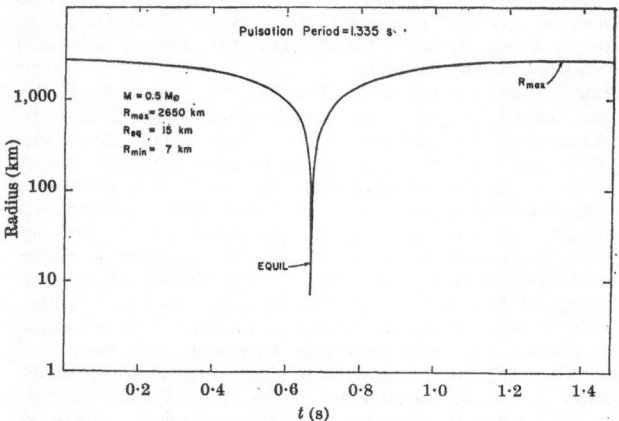

Fig. 2. Outer radius of a neutron star as a function of time.

Gyroradiation might also be a contributing source. For 100 Hz, the appropriate field strength is 36 gauss, but for non-relativistic electrons, the decay time for the electron kinetic energy[18] in such a field is roughly 5×10^5 s. Thus, while it might be possible to get enough energy stored as electron kinetic energy, many more orders of magnitude of participating electrons would be required for the necessary emissivity. Gyroradiation can still contribute, however, because bursts with fast rise-times might result from stimulated "maser" emission operating in layers where the electron distribution is non-thermal. This would give nearly monochromatic emission, the bandwidth being determined by inhomogeneities in the magnetic field and by the electron relativity parameter. If the pulsation amplitude is large, the two radio bursts per stellar oscillation may have different characteristics. On the other hand, it is likely that the amplitudes are small, for the calculations described in the previous section show that collapse occurs when the amplitudes are too large.

We thank W. D. Arnett, A. G. W. Cameron, H. van Horn, A. Lapidus, W. Quirk, J. J. Rickard, E. E. Salpeter, A. Schindler and R. Stothers for helpful discussions. This work was supported in part by NAS-NRC research associateships sponsored by the US National Aeronautics and Space Administration. It is based in part on a lecture by W. J. Cocke at the University of Sussex in June 1968, and in part on a paper by J. M. Cohen at the Pulsar Conference, May 20–21, 1968, New York.

Received July 22; revised July 25, 1968.

[1] Cameron, A. G. W., and Maran, S. P., *Sky and Telescope*, **36**, 4 (1968).
[2] Faulkner, J., and Gribbin, J., *Nature*, **218**, 734 (1968), (Paper 28).
[3] Skilling, J., *Nature*, **218**, 531 (1968), (Paper 30).
[4] Chandrasekhar, S., *Stellar Structure* (University of Chicago Press, Chicago, 1938).
[5] Hamada, T., and Salpeter, E., *Ap. J.*, **134**, 638 (1961).
[6] Skilling, J., *Nature*, **218**, 928 (1968), (Paper 31).
[7] Christy, R. F., *Rev. Mod. Phys.*, **36**, 555 (1964).
[8] Colgate, S. A., and White, R. H., *Ap. J.*, **143**, 626 (1966).
[9] Arnett, W. D., *Canad. J. Phys.*, **44**, 2553 (1966).
[10] Cameron, A. G. W., *Stellar Evolution, Nuclear Astrophysics and Nucleogenesis* (Chalk River, Ontario, 1957).
[11] Christy, R. F., *Ann. Rev. Astro. Astrophys.*, **4**, 383 (1966).
[12] Hoyle, F., and Narlikar, J., *Nature*, **218**, 123 (1968), (Paper 27).
[13] Alfven, H., and Falthammar, C. G., *Cosmical Electrodynamics*, second ed., 161 (Oxford University Press, 1963).
[14] Wild, J. P., Smerd, S. F., and Weiss, A. A., *Ann. Rev. Astro. Astrophys.*, **1**, 360 (1963).
[15] Kundu, M. R., and Chitre, S. M., *Nature*, **218**, 1037 (1968), (Paper 36).
[16] Alfven, H., and Falthammar, C. G., *Cosmical Electrodynamics*, second ed., 69 (Oxford University Press, 1963).
[17] Maran, S. P., and Cameron, A. G. W., *Physics Today* (in the press).
[18] Wild, J. P., Smerd, S. F., and Weiss, A. A., *Ann. Rev. Astro. Astrophys.*, **1**, 354 (1963).

39. Possible Model for a Rapidly Pulsating Radio Source

by
JEREMIAH OSTRIKER
Princeton University Observatory,
Princeton, New Jersey

RECENTLY, the Cambridge radio astronomy group has described observations of a rapidly pulsating radio source[1]. Their observations can be summarized as follows. (a) The signal observed at 81·5 MHz has a period of 1·337 s, the phase being constant to one part in 10^7. The amplitude, however, varies irregularly, the signal appearing and disappearing for periods of a few minutes. Each pulse has a length of $\sim 0 \cdot 3$ s. (b) From the absence of proper motion and a consideration of the dispersive effects of transmission through the interstellar plasma the distance to the source is limited to the range 10^3 AU $< d < 65$ pc. (c) Variations of phase can be accounted for by the Earth's orbital motion without the need for additional correction for the motion of the source. (d) No visible object brighter than twelfth magnitude is apparent in the radio error rectangle. (e) From the signal width and rate of sweep the source size must be smaller than $4 \cdot 8 \times 10^8$ cm.

The authors[1] tentatively suggested that the radio signals come from a pulsating neutron star or white dwarf. The calculations of Meltzer and Thorne[2], however, show that stable (as opposed to unstable or metastable) neutron stars have central densities in the range $2 \cdot 7 \times 10^{13} \leqslant \rho_c \leqslant 6 \cdot 0 \times 10^{15}$ and fundamental modes of radial oscillation with periods in the range $5 \times 10^{-2} \geqslant T \geqslant 8 \times 10^{-4}$. Overtones of the fundamental modes have shorter periods still, so pulsating neutron stars seem to be an unlikely origin for the radio signals. On the other hand, white dwarf radial oscillations always have periods[2] exceeding 8 s. The equilibrium models, however, assume that the nuclear matter of the white dwarfs has been catalysed to the end point of thermonuclear evolution. A more realistic equation of state derived by Salpeter[3] produces equilibrium

Table 1. MODEL OF A RAPIDLY ROTATING WHITE DWARF

M	$2 \cdot 26\ M \odot$
J	$3 \cdot 46 \times 10^{50}$ g cm^2 s^{-1}
$R_{\frac{1}{2}}$	$1 \cdot 35 \times 10^8$ cm
R_1	$4 \cdot 44 \times 10^8$ cm
ρ_c	$3 \cdot 90 \times 10^8$ g cm^{-3}
R_e/R_p	$2 \cdot 14$
E	$-1 \cdot 21 \times 10^{51}$ ergs
P_0	$1 \cdot 25$ s
$P_{\frac{1}{2}}$	$1 \cdot 61$ s
P_1	$6 \cdot 50$ s

models of higher density[4] and consequently shorter period. In any case, observations (a) are very difficult to understand if the source is a pulsating star. With the large energy that would be stored in a stellar pulsation it is not easy to see how the amplitude could be so variable. If surface pulsations are adopted it is easier to understand how the signal could appear and disappear at irregular intervals but very difficult to understand why the phase should be constant over successive reappearances. Because it has been suggested that the radio source may be associated with a planet, it is perhaps worth noting that from (b) and (d) the associated star must be faint (fainter than eighth magnitude) but is likely to be further from the planet than 1 AU (c).

We require an intrinsically faint (b, d), small (e) star having a natural period of ~ 1 s (a). For the reasons mentioned it seems unlikely that the periodic behaviour is associated with a pulsation period. Periodic behaviour is also found astronomically in two other cases, orbital motion and rotation, both of which can cause periodic variations in apparent magnitude.

We can estimate the shortest possible period for orbiting ordinary stars if we consider a congruent pair of white dwarfs which are assumed to be undistorted by tidal or rotational forces and which are in a grazing circular orbit about one another. For stars of mass M and radius R the period P must satisfy the inequality

$$P > 4\pi \left(\frac{R^3}{GM}\right)^{1/2} > 1.7\,\text{s}$$

Tidal and rotational distortion will increase this limit by about 50 per cent. The numerical result is obtained by using the mass and radius ($1.396\,M_\odot$, $0.216 \times 10^{-2}\,R_\odot$) derived[4] for a carbon white dwarf on the verge of implosion due to inverse β decay. For a pair of neutron stars the desired orbital period can be obtained, but gravitational radiation caused by the varying quadrupole moment[5] would produce a continuous rate of frequency shift equal to

$$\left(\frac{\Delta\omega}{\omega}\right) = \frac{96}{5} \frac{\varepsilon}{(1+\varepsilon)^{1/3}} \frac{(GM)^{5/3}}{c^5} \omega^{8/3}\Delta t$$

Stars having masses M and εM and circular orbits about their mutual centre of mass have been assumed. For stable neutron stars, $M > 0.18\,M_\odot$, $4.0 > \varepsilon > 0.25$, so that the expected change in frequency would be

$$\frac{\Delta\omega}{\omega} > 2 \times 10^{-3}\,\text{day}^{-1}$$

a possibility which can be ruled out by the observations.

Finally, we consider rotation. The angular velocity at the equator of a rapidly rotating, condensed, star can be estimated by the Keplerian relation

$$\Omega = \left(\frac{GM}{R^3}\right)^{1/2}$$

which, if the period ($P = 2\pi/\Omega$) is to be the order of 1 s, gives a "density" ($3M/4\pi R^3$) of the order of 10^8 g/cm^{-3}. This density is characteristic of the matter in the more massive white dwarf stars. Recently[6,7] various sequences of models of rapidly, differentially rotating white dwarfs have been constructed with masses both above and below the mass limit[8] ($1.44\,M_\odot$) for non-rotating white dwarfs. The models with rotation periods as short as 1 s have masses in the range 1–3 $M_\odot$. In Table 1 we summarize the properties of a 2.26 $M_\odot$ model with angular momentum $J = 3.46 \times 10^{50}$ g cm^2 s^{-1} (characteristic of the main-sequence star having the same mass) and the same angular momentum distribution as in a uniformly rotating uniform sphere.

In Table 1, R is the radius, P the period of rotation, ρ_c the central density and E the binding energy. The subscripts "0", "$\frac{1}{2}$" or "1" refer to quantities measured on a cylinder containing that fraction of the mass; subscripts "e" and "p" refer to the equator and the pole. We note, in connexion with observation (e), that the star is smaller than 5×10^8 cm. This very rapidly rotating star ($v \sim 5,000$ km s^{-1}) might be expected to be in danger of fission caused by an instability to a nonaxisymmetric mode. The lowest ($m = 1$, 2) modes have been checked by Tassoul and myself and the star is found to be stable in the absence of viscosity but unstable, with an e-folding time of $\sim 10^{10}$ yr, when viscous forces are included; that is, the model is stable.

We can imagine an active region on such a star emitting radiation similar to type II or III solar bursts, each burst lasting the order of a few minutes. During this time radio energy is emitted in the direction of the Earth as a set of pulses, one per period, the period being characteristic of the latitude of the active region. For the time of emission to be as short as inferred from the observations (0.016 s) it is necessary that the radiation be emitted in a cone of width $\sim 1°$. A subsequent burst from the same active region would appear on the observer's record as a new set of pulses with amplitude unrelated to the first set but exactly in phase with them. If active regions appear at other points at the same latitude, sudden phase shifts would be observed caused by differential rotation; a shift to active regions at other latitudes would lead to changes in both phase and period.

The details of the model presented here are clearly very tentative. The major point to be made is that rotation provides perhaps the best explanation for the periodicity of the new radio sources, and that rapidly rotating white dwarfs have rotation periods in the experimentally observed range.

This work was supported in part by grants from the Air Force Office of Scientific Research and the US National Science Foundation.

Received March 18, 1968.

[1] Hewish, A., Bell, S. J., Pilkington, J. D. H., Scott, P. F., and Collins, R. A., *Nature*, 217, 709 (1968), (Paper 1).
[2] Meltzer, D. W., and Thorne, K. S., *Astrophys. J.*, 145, 514 (1966).
[3] Salpeter, E. E., *Astrophys. J.*, 134, 669 (1961).
[4] Hamada, T., and Salpeter, E. E., *Astrophys. J.*, 134, 683 (1961).
[5] Landau, L., and Lifshitz, E., *The Classical Theory of Fields* (Addison-Wesley, 1951).
[6] Ostriker, J. P., Bodenheimer, P., and Lynden-Bell, D., *Phys. Rev. Lett.*, 17, 816 (1966).
[7] Ostriker, J. P., and Bodenheimer, P., *Astrophys. J.*, 151, 1089 (1968).
[8] Chandrasekhar, S., *Astrophys. J.*, 74, 81 (1931).

40. Rotating Neutron Stars as the Origin of the Pulsating Radio Sources

by

T. GOLD

Center for Radiophysics and Space Research,
Cornell University,
Ithaca, New York

THE case that neutron stars are responsible for the recently discovered pulsating radio sources[1-6] appears to be a strong one. No other theoretically known astronomical object would possess such short and accurate periodicities as those observed, ranging from 1.33 to 0.25 s. Higher harmonics of a lower fundamental frequency that may be possessed by a white dwarf have been mentioned; but the detailed fine structure of several short pulses repeating in each repetition cycle makes any such explanation very unlikely. Since the distances are known approximately from interstellar dispersion of the different radio frequencies, it is clear that the emission per unit emitting volume must be very high; the size of the region emitting any one pulse can, after all, not be much larger than the distance light travels in the few milliseconds that represent the lengths of the individual pulses. No such concentrations of energy can be visualized except in the presence of an intense gravitational field.

The great precision of the constancy of the intrinsic period also suggests that we are dealing with a massive object, rather than merely with some plasma physical configuration. Accuracies of one part in 10^8 belong to the realm of celestial mechanics of massive objects, rather than to that of plasma physics.

It is a consequence of the virial theorem that the lowest mode of oscillation of a star must always have a period which is of the same order of magnitude as the period of the fastest rotation it may possess without rupture. The range of 1·5 s to 0·25 s represents periods that are all longer than the periods of the lowest modes of neutron stars. They would all be periods in which a neutron star could rotate without excessive flattening. It is doubtful that the fundamental frequency of pulsation of a neutron star could ever be so long (ref. 7 and unpublished work of A. G. W. Cameron). If the rotation period dictates the repetition rate, the fine structure of the observed pulses would represent directional beams rotating like a lighthouse beacon. The different types of fine structure observed in the different sources would then have to be attributed to the particular asymmetries of each star (the "sunspots", perhaps). In such a model, time variations in the intensity of emission will have no effect on the precise phase in the repetition period where each pulse appears; and this is indeed a striking observational fact. A fine structure of pulses could be generated within the repetition period, depending only on the distribution of emission regions around the circumference of the star. Similarly, a fine structure in polarization may be generated, for each region may produce a different polarization or be overlaid by a different Faraday-rotating medium. A single pulsating region, on the other hand, could scarcely generate a repetitive fine structure in polarization as seems to have been observed now[8].

There are as yet not really enough clues to identify the mechanism of radio emission. It could be a process deriving its energy from some source of internal energy of the star, and thus as difficult to analyse as solar activity. But there is another possibility, namely, that the emission derives its energy from the rotational energy of the star (very likely the principal remaining energy source), and is a result of relativistic effects in a co-rotating magnetosphere.

In the vicinity of a rotating star possessing a magnetic field there would normally be a co-rotating magnetosphere. Beyond some distance, external influences would dominate, and co-rotation would cease. In the case of a fast rotating neutron star with strong surface fields, the distance out to which co-rotation would be enforced may well be close to that at which co-rotation would imply motion at the speed of light. The mechanism by which the plasma will be restrained from reaching the velocity of light will be that of radiation of the relativistically moving plasma, creating a radiation reaction adequate to overcome the magnetic force. The properties of such a relativistic magnetosphere have not yet been explored, and indeed our understanding of relativistic magnetohydrodynamics is very limited. In the present case the coupling to the electromagnetic radiation field would assume a major role in the bulk dynamical behaviour of the magnetosphere.

The evidence so far shows that pulses occupy about 1/30 of the time of each repetition period. This limits the region responsible to dimensions of the order of 1/30 of the circumference of the "velocity of light circle". In the radial direction equally, dimensions must be small; one would suspect small enough to make the pulse rise-times comparable with or larger than the flight time of light across the region that is responsible. This would imply that the radiation emanates from the plasma that is moving within 1 per cent of the velocity of light. That is the region of velocity where radiation effects would in any case be expected to become important.

The axial asymmetry that is implied needs further comment. A magnetic field of a neutron star may well have a strength of 10^{12} gauss at the surface of the 10 km object. At the "velocity of light circle", the circumference of which for the observed periods would range from 4×10^{10} to 0.75×10^{10} cm, such a field will be down to values of the order of 10^3–10^4 gauss (decreasing with distance slower than the inverse cube law of an undisturbed dipole field. A field pulled out radially by the stress of the centrifugal force of a whirling plasma would decay as an inverse square law with radius). Asymmetries in the radiation could arise either through the field or the plasma content being non-axially symmetric. A skew and non-dipole field may well result from the explosive event that gave rise to the neutron star; and the access to plasma of certain tubes of force may be dependent on surface inhomogeneities of the star where sufficiently hot or energetic plasma can be produced to lift itself away from the intense gravitational field (10–100 MeV for protons; much less for space charge neutralized electron-positron beams).

The observed distribution of amplitudes of pulses makes it very unlikely that a modulation mechanism can be responsible for the variability (unpublished results of P. A. G. Scheuer and observations made at Cornell's Arecibo Ionospheric Observatory) but rather the effect has to be understood in a variability of the emission mechanism. In that case the observed very sharp dependence of the instantaneous intensity on frequency (1 MHz change in the observation band gives a substantially different pulse amplitude) represents a very narrow-band emission mechanism, much narrower than synchrotron emission, for example. A coherent mechanism is then indicated, as is also necessary to account for the intensity of the emission per unit area that can be estimated from the lengths of the sub-pulses. Such a coherent mechanism would represent non-uniform static configurations of charges in the relativistically rotating region. Non-uniform distributions at rest in a magnetic field are more readily set up and maintained than in the case of high individual speeds of charges, and thus the configuration discussed here may be particularly favourable for the generation of a coherent radiation mechanism.

If this basic picture is the correct one it may be possible to find a slight, but steady, slowing down of the observed repetition frequencies. Also, one would then suspect that more sources exist with higher rather than lower repetition frequency, because the rotation rates of neutron stars are capable of going up to more than 100/s, and the observed periods would seem to represent the slow end of the distribution.

Work in this subject at Cornell is supported by a contract from the US Office of Naval Research.

Received May 20, 1968.

[1] Hewish, A., Bell, S. J., Pilkington, J. D. H., Scott, P. F., and Collins, R. A., Nature, 217, 709 (1968), (Paper 1).
[2] Pilkington, J. D. H., Hewish, A., Bell, S. J., and Cole, T. W., Nature, 218, 126 (1968), (Paper 2).
[3] Drake, F. D., Gundermann, E. J., Jauncey, D. L., Comella, J. M., Zeissig, G. A., and Craft, jun., H. D., Science, 160, 503 (1968).
[4] Drake, F. D., Science (in the press).
[5] Drake, F. D., and Craft, jun., H. D., Science, 160, 758 (1968).
[6] Tanenbaum, B. S., Zeissig, G. A., and Drake, F. D., Science (in the press).
[7] Thorne, K. S., and Ipser, J. R., Ap. J. (in the press).
[8] Lyne, A. G., and Smith, F. G., Nature, 218, 124 (1968), (Paper 7).

41. Rotating Neutron Stars, Pulsars and Supernova Remnants

by

F. PACINI*

Center for Radiophysics and Space Research,
Cornell University,
Ithaca, New York

I SHALL discuss here some problems connected with theories linking the pulsars to the rotation of neutron stars (ref. 1 and a preprint by L. Woltjer). Because neutron stars can be formed during a supernova explosion, their rotation could be coupled with the surrounding gaseous remnant[2,3]: the following considerations will therefore also refer to the problem of the activity observed in the Crab Nebula and similar objects.

Gold (private communication) has noted that the direct detection of a pulsar in the Crab Nebula would be made very difficult by the dispersion inside the nebula itself. It is also obvious that a pulsed emission will not result if the basic cause of the pulses (for instance, the stellar rotation) has a time scale shorter than the width of the individual pulses. A rotating neutron star could then be present, say, in the Crab Nebula even if there is no evidence for a pulsating radio source in this part of the sky (Drake, private communication). Long term brightness variations similar to the ones found in the pulsars could still be detectable: a search for this kind of variability in the compact low frequency source in the Crab is needed.

Following my earlier paper[3] and more recent remarks by several authors (ref. 1 and preprints by L. Woltjer and by B. J. Eastlund), I shall consider a neutron star which has a dipole magnetic field and rotates about an axis different from that of the magnetic field. The oblique rotator configuration seems common among magnetic stars where the angle between the field and the rotation axes tends to be close to 90° (ref. 4). Even if the two axes coincided in the original star, the mass loss during the supernova explosion would certainly not be perfectly spherically symmetric, especially if large magnetic fields are present. As a consequence, the mutual inclination of the axes in a newly born neutron star is likely to be almost arbitrary. Spitzer[5] has noted that an oblique rotator configuration cannot persist indefinitely because the body is rotating about an axis which is not a principal axis of inertia. Dissipation mechanisms in the form of hydromagnetic waves can be important close to the surface of the star and eventually result in an acceleration of particles. Remarkable evidence for this kind of activity close to the magnetic poles of some A_p stars comes from the distribution of peculiar elements in these stars (ref. 4 and a preprint by L. Woltjer). The possibility of a connexion between this kind of hydromagnetic activity in an oblique rotator and the activity observed in some supernova remnants is worth mentioning.

I shall not attempt to evaluate a lifetime for the oblique rotator configuration, as the problem is extremely difficult. There is, however, a difficulty of a different kind associated with the existence of the magnetic field itself in an old neutron star. The decay time of the field is proportional to the product σR^2, where σ is the electrical conductivity and R is the radius of the star. An order of magnitude for σ can be roughly estimated by putting

$$\sigma = \frac{e^2}{m_e v S} \tag{1}$$

* On leave from Laboratorio d'Astrofisica, Frascati, Italy.

where v is the electron velocity and S is the Coulomb cross-section. For a relativistic electron gas we take $v = c$ and S of the order of the classical value $\pi r_e^2 = 10^{-24}$ cm². The electrical conductivity turns out to be $\sigma \sim 10^{22}$ s⁻¹, that is, about 10^4–10^5 times larger than in the Sun. For a neutron star we have $R \simeq 10^6$ cm $\simeq 10^{-5} R_\odot$, and for the Sun the decay time is about 10^{10} yr, so it is at least uncertain whether the fossil magnetic field of a neutron star can last longer than about 10^4–10^5 yr. It has to be emphasized that this estimate does not take into account the gas degeneracy and, even more important, the possibility of a superconductive behaviour of the neutron star matter in some density range. A complete investigation of the problem is certainly desirable.

I shall now describe the electromagnetic field around a neutron star rotating with an angular velocity ω in a quasi-vacuum. By quasi-vacuum we mean that the field of the star cannot be compensated by the currents induced in the surrounding medium. The corresponding density limit can be evaluated by noting that the maximum current density in the circumstellar gas is $j_{max} = n_e e c$. The Maxwell equation curl $\mathbf{H} = (4\pi/c) \mathbf{j}$ then gives the maximum induced field. If we take curl $H \sim 1/r$ (r is of the order of the size of the system) we obtain $H_{max} \simeq 4\pi n_e r$. The field of the star, or variations of the field of the order of the field itself, cannot be compensated by the induced currents if $n_e < H/(4\pi e r)$. Assuming $H \simeq 10^{10}$ gauss and $r \simeq R \simeq 10^6$ cm the critical electron density is about 10^{13} cm⁻³. This limit is certainly not very stringent and is likely to be violated only soon after the birth of the neutron star. The assumption of a quasi-vacuum then seems legitimate near a neutron star having a very strong magnetic field.

Let α be the angle between the magnetic and the rotation axes. If $\alpha \neq 0$ retardation effects become important for the field at a distance r such that $\omega r \sim c$: for $\omega r \gg c$ the field becomes a wave progressing outwards. As noted[3] the oblique rotator model leads to the release of the rotational energy of the star by the radiation of electromagnetic waves having the same frequency as the star rotation.

Deutsch has given a complete description of the field of a rotating sphere[6]. Following this work we shall introduce an inertial reference system (r, θ, φ) having its origin in the centre of the star and axis ω. If we assume that the internal field is frozen into the star, its components can be written

$$H(r, \theta, \varphi) = \mathbf{e}_r H_r(r, \theta, \lambda) + \mathbf{e}_\theta H_\theta(r, \theta, \lambda) + \mathbf{e}_\varphi H_\varphi(r, \theta, \lambda) \tag{2}$$

where λ is an azimuthal coordinate measured from a meridian fixed in the star. In the same reference system the internal electric field is given by

$$\mathbf{E} = -\mu_0(\omega \times \mathbf{r}) \times \mathbf{H} \tag{3}$$

If we impose the continuity conditions at the surface of the star for the field components, the external field is fully determined by the Maxwell equations.

Let H_0 be the surface value of the magnetic field and a be the radius of the star: the general expression for the external electromagnetic field turns out to be very complicated[6] and I shall only give its components in the near region $r \ll (c/\omega)$ and in the far-wave region $r \gg (c/\omega)$.

(a) In the near region the field rotates with the star and depends on the azimuthal coordinate λ which is measured from a meridian fixed in the star. The components are

$$H_r = H_0 \left(\frac{a}{r}\right)^3 [\cos \alpha \cos \theta + \sin \alpha \sin \theta \cos \lambda]$$

$$H_\theta = \frac{1}{2} H_0 \left(\frac{a}{r}\right)^3 [\cos \alpha \sin \theta - \sin \alpha \cos \theta \cos \lambda]$$

$$H_\varphi = \frac{1}{2} H_0 \left(\frac{a}{r}\right)^3 \sin \alpha \sin \lambda \qquad (4)$$

$$E_r = - \frac{1}{4} \omega a \mu_0 H_0 \left(\frac{a}{r}\right)^4 [\cos \alpha (3 \cos 2\theta + 1) + 3 \sin \alpha \sin 2\theta \cos \lambda]$$

$$E_\theta = - \frac{1}{2} \omega a \mu_0 H_0 \left(\frac{a}{r}\right)^2 \left[\frac{a^2}{r^2} \cos \alpha \sin 2\theta + \sin \alpha \left(1 - \frac{a^2}{r^2} \cos 2\theta\right) \cos \lambda\right] \qquad (5)$$

$$E_\varphi = \frac{1}{2} \omega a \mu_0 H_0 \left(\frac{a}{r}\right)^2 \left(1 - \frac{a^2}{r^2}\right) \sin \alpha \cos \theta \sin \lambda$$

(b) Far-wave zone $r \gg (c/\omega)$; in this case the retardation effects dominate and we obtain

$$H_r = \frac{\omega}{c} H_0 a^3 \frac{1}{r^2} \sin \alpha \sin \theta \sin \left[\omega \left(\frac{r}{c} - t\right) + \varphi\right]$$

$$H_\theta = \frac{1}{2} \frac{\omega^2}{c^2} a^3 H_0 \frac{1}{r} \sin \alpha \cos \theta \cos \left[\omega \left(\frac{r}{c} - t\right) + \varphi\right] \qquad (6)$$

$$H_\varphi = - \frac{1}{2} \frac{\omega^2}{c^2} a^3 H_0 \frac{1}{r} \sin \alpha \sin \left[\omega \left(\frac{r}{c} - t\right) + \varphi\right]$$

$$E_r = 0$$

$$E_\theta = - \frac{1}{2} \frac{\omega^2 \mu_0}{c} a^3 H_0 \frac{1}{r} \sin \alpha \sin \left[\omega \left(\frac{r}{c} - t\right) + \varphi\right] \qquad (7)$$

$$E_\varphi = - \frac{1}{2} \frac{\omega^2 \mu_0}{c} a^3 H_0 \frac{1}{r} \sin \alpha \cos \theta \cos \left[\omega \left(\frac{r}{c} - t\right) + \varphi\right]$$

If the magnetic and rotation axes coincide, there is no radiation field; the electromagnetic field external to the star can be obtained simply by putting $\alpha = 0$ in the set of equations (4) and (5).

By using equations (6) and (7) we can obtain the Poynting vector and evaluate the rate at which the rotational energy W and the angular momentum L are radiated away from the star. The result is

$$\frac{dW}{dt} = - \frac{2\pi \omega^4 \mu_0}{3 c^3} a^6 H_0^2 \sin^2 \alpha \qquad (8)$$

and

$$\frac{dL}{dt} = \frac{1}{\omega} \frac{dW}{dt} \qquad (9)$$

Once the electromagnetic waves are emitted and start to propagate, they will be reflected by the circumstellar gas if the plasma frequency exceeds the radiation frequency[3]. In our case this certainly occurs for every conceivable value of ω and gas density, and the electromagnetic waves will be unable to reach us. Generation of high energy particles can, however, be expected in the region where the waves are reflected and cause a rapid compression of the nebular gas. Because the radiation pattern of the star is directional, it becomes possible to speculate whether a pulsed radio source arises because of this periodical gas compression and consequent acceleration of particles. From an energy point of view no difficulty arises: the period of pulsar CP 1919 corresponds to $\omega = 4 \cdot 7$ s^{-1} and if we assume $H \sin \alpha \simeq 10^{10-11}$ gauss the energy output is 10^{27}-10^{29} ergs s^{-1} in agreement with the observational requirements. On the other hand, for a newly formed neutron star, rotational frequencies 100 or even 1,000 times larger are perfectly possible and the energy output corresponding to fields of 10^{11} gauss can be as high as 10^{41} erg s^{-1}. The high rotational velocity itself could wash out the pulse structure and lead to a continuum emission: this situation possibly corresponds to what is observed in the Crab Nebula.

Finally, in this model a pulsating radio source would show a relatively slow change in period. From equation (8) we can obtain an expression for the rotational energy $W = (1/2) k M a^2 \omega^2$ left at a certain time t as a function of the original rotational energy. We have

$$\frac{W_0}{W(t)} = 1 + \frac{4\pi a^4}{3 c^3 k M} \omega_0^2 (H_0 \sin \alpha)^2 t \qquad (10)$$

For a neutron star we can assume $k = 0 \cdot 2$, $a = 10^6$ cm and $M = 2 \times 10^{33}$ g. As $W \propto T^{-2}$ we have for the periods

$$T^2 = T_0^2 (1 + 1 \cdot 2 \times 10^{-32} \omega_0^2 H_0^2 \sin^2 \alpha \, t_{yr}) \qquad (11)$$

where T_0 is the initial rotation period. We can then put a limit to the product $H \sin \alpha$ for the pulsar CP 1919 because we know that the fractional change in period $\Delta T/T_0$ over 1 yr cannot exceed 10^{-7}. This gives $H_0 \sin \alpha < 10^{12}$ gauss.

It was my chief concern to discuss some possibilities which arise when the oblique configuration for the magnetic field of a rotating neutron star is assumed. In particular, I wanted to show that the model leads to a release of rotational energy from the star at a rate which can agree quantitatively with the requirements either of the pulsars or of objects such as the Crab Nebula.

I emphasize, however, that the consideration of the circumstellar plasma is going to be of prime importance in several respects. This plasma is the place where the radiation we actually observe originates: the rotating neutron star only provides the energy source and the timing of the basic excitation (periodic compression of the plasma by the low frequency electromagnetic waves). Hoyle, Narlikar and Wheeler[7] have noted that the scale height for a neutron star is very small and therefore the density of the gas that may exist outside the star is also small. Furthermore, the pressure exerted by the radiation is enormous and any residuum of gas near the star would be swept outwards. The neutron star itself will therefore be surrounded by the vacuum (at least as long as radiation pressure dominates over gravitational accretion) but, farther away, the electromagnetic waves will be reflected by the interstellar gas. The equations (8) to (11) assume radiation in a vacuum: whether or not the same loss of rotational energy occurs in real life seems to hinge on what happens to the energy and angular momentum of the electromagnetic waves when they interact with the plasma.

Finally, we note that the same model could also apply to a very rapidly rotating white dwarf. In this case, however, magnetic fields of 10^{10} gauss seem very unlikely and it would be more difficult to achieve the right conditions in the circumstellar gas.

I thank Professor T. Gold and Professor E. E. Salpeter for many discussions and critical remarks. This work was supported by the US Office of Naval Research.

Received June 11, 1968.

[1] Gold, T., Nature, 218, 731 (1968), (Paper 40).
[2] Wheeler, J. A., Ann. Rev. Astro. Astrophys., 4 (1966).
[3] Pacini, F., Nature, 216, 567 (1967).
[4] Preston, G. W., Ap. J., 150, 547 (1967).
[5] Spitzer, L., Electromagnetic Phenomena in Cosmical Physics (IAU Symposium) (1958).
[6] Deutsch, A., Annales d'Astrophysique, 18, 1 (1955).
[7] Hoyle, F., Narlikar, J., and Wheeler, J. A., Nature, 203, 914 (1964).

42. Rapidly Pulsing Radio Sources

by

WILLIAM C. SASLAW
Department of Applied Mathematics and
Theoretical Physics,
Silver Street, Cambridge

JOHN FAULKNER
PETER A. STRITTMATTER
Institute of Theoretical Astronomy,
Madingley Road, Cambridge

OBSERVATIONS of a new type of rapidly varying radio source have recently been reported[1]. The principal characteristics of the four sources so far discovered may be summarized as follows—sharp pulses of radiation are received, separated by comparatively long intervals which themselves are remarkably constant (with variations of $\lesssim 1$ part in 2×10^7 during the past 3 or 4 months). The pulse duration is no more than ~ 0.016 s, while the period is 1.3372795 ± 0.0000020 s (ref. 1). We understand that the other objects show similar sharp pulses at intervals of the order of 1 s, which, although not identical, strongly suggest that a new type of physical system has been discovered. Hewish *et al.* propose that the objects may be pulsating white dwarfs or neutron stars. Although this may prove to be correct, there are certain difficulties associated with the constancy of the observed period. We suggest an alternative explanation for the observed phenomena.

Gravitational Lens Hypothesis

An obvious method of achieving astronomical regularity is that associated with motion in a binary system. We therefore suggest that the radiation peaks are caused by the nearest member of a binary pair actingas as a gravitational lens on the radiation of its more distant companion. The observed interval between emission bursts for objects of equal mass would be one-half the orbital period. It might be thought that such occurrences would necessarily have a negligible probability. Remarkably, this is not the case.

We first embark on an approximate illustrative analysis, to be followed by more detailed computations. Consider the motion of two stars of equal mass $M \sim 0.5 \, M_{\odot}$ and with an orbital period of 3 s. This corresponds to a separation $a \approx 3 \times 10^8$ cm and orbital velocities $\sim 10^{-2} \, c$. Clearly, it follows from the small separation that the stars must be very compact white dwarfs or neutron stars. This requirement is also necessary for the gravitational lens mechanism to be effective.

Liebes[2] has given a theoretical discussion of the gravitational lens and we shall follow his notation throughout. Fig. 1 illustrates the configuration. O, D and R denote the source, deflector and receiver, respectively. $2\varphi_O$ and $2\varphi_D$ denote the undistorted angular diameters of O and D, if they could be seen by the receiver. The semi-apex angle of the "cone of inversion" is given by

$$\theta_0 = \left(\frac{4 \, G \, M_D}{(1 + l_D/l_{OD}) \, l_D \, c^2} \right)^{1/2}$$

$$\approx \left(\frac{4 \, G \, M_D}{c^2 \, l_D} \cdot \frac{l_{OD}}{l_D} \right)^{1/2} \tag{1}$$

because in our case $l_D \gg l_{OD}$. M_D denotes the mass of

the deflector. As the source moves behind the deflector the receiver sees two images I_1 and I_2 on either side of the cone of inversion. These images merge to produce an annulus round the cone of inversion in the unlikely case when O, D and R are perfectly aligned. Because the brightness remains the same along a light path, the amplification factor A is given by the ratio of the total solid angle subtended by the deflected images to the undistorted solid angle of the source. Thus Liebes obtains

$$A \approx 1, \quad \alpha > \theta_0$$
$$\approx \theta_0/\alpha, \quad \alpha < \theta_0 \tag{2}$$

with a maximum value

$$A_{\max} \approx 2\theta_0/\varphi_0 \tag{3}$$

in the case of perfect alignment. These results are subject to the conditions that all deflexion angles are small and that the geometrical cross-section of the deflector is too small to block much radiation. This last condition implies that

$$\frac{\theta_0}{\varphi_D} = \left(\frac{4 \, G \, M_D}{c^2 \, r_D} \right)^{1/2} \left(\frac{l_{OD}}{r_D} \right)^{1/2} \gg 1 \tag{4}$$

where r_D is the deflector radius. Condition (4) can be satisfied only when $l_{OD}/r_D \gg 1$, because the first term in brackets is essentially the Schwarzschild parameter.

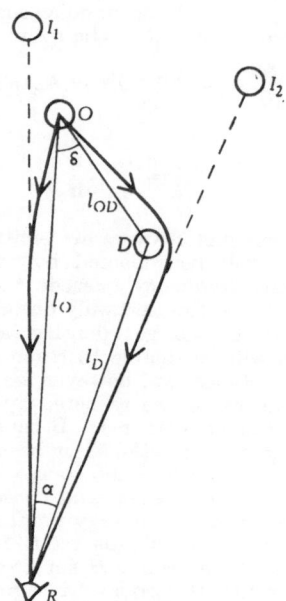

Fig. 1. **Geometric configuration and typical light paths for gravitational lens.**

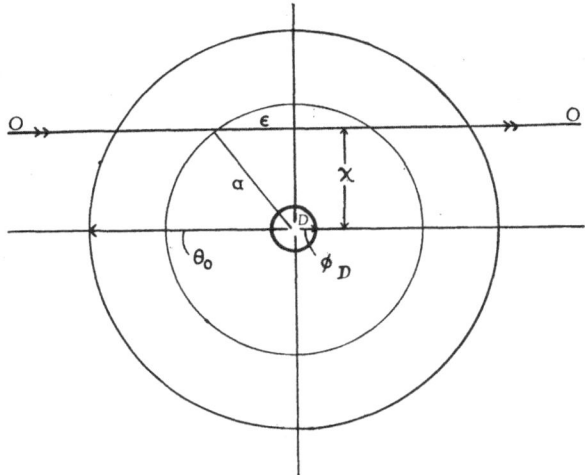

Fig. 2. Path of centre of source $(O - O)$ compared with cone of inversion.

With the estimated value of l_{OD} $(\equiv a)$ given here, this implies that the deflector cannot be a white dwarf and must therefore presumably be a neutron star. Similarly, to obtain substantial amplification, we require that

$$A_{max} \approx \frac{2\theta_0}{\varphi_0} = 4\left(\frac{GM_D}{c^2 r_0}\right)^{1/2}\left(\frac{l_{OD}}{r_0}\right)^{1/2} \gg 1 \qquad (5)$$

which again demands that the source be a neutron star. We are led therefore to a binary system consisting of two neutron stars, each acting as a gravitational lens for the other.

In the following discussion we will consider only those cases in which reasonable amplification $(A \gtrsim 2)$ is obtained. The receiver R will not usually be in the orbital plane of the stars and we define χ to be the angle of closest approach as seen by R. The situation is illustrated in Fig. 2, in which the path of O is taken to have negligible curvature within the cone of inversion. This is a good approximation provided $(4GM_D/l_{OD}c^2)^{1/2} \ll 1$, which is easily satisfied in the present case. Clearly the largest amplification A_m for a given path is

$$A_m \approx \theta_0/\chi = \frac{\varphi_0}{2\chi_0} \cdot A_{max} \qquad (6)$$

The angle δ_A $(\angle DOR)$ at a given amplification is then given by

$$\delta_A = (l_D/l_{OD})\alpha = l_D\theta_0/A \qquad (7)$$
$$= (1/A)\,(4GM_D/l_{OD}c^2)^{1/2}$$

The time interval between points of given amplification in a single pulse is thus given by

$$T_A = (\delta_A/\pi)\xi P \qquad (8)$$

where

$$\xi = A\varepsilon/\theta_0 = (1 - A^2\chi^2/\theta_0^2)^{1/2} \qquad (9)$$

provided, of course, that the required amplification is reached, and that $A \lesssim \theta_0/\varphi_0$. If we demand, for example, an amplification of $A = 3$ and use the approximate values of M_D and l_{OD} assumed here, we obtain $\delta_3 \approx 10^{-2}$ and $T_3 \lesssim (\delta_3/\pi)\,P \simeq 10^{-2}$ s. This time certainly satisfies the condition $T \lesssim 0.016$ s given by Hewish *et al.* and suggests that, with entirely reasonable estimates of M, the observed characteristics of the signal can be reproduced.

The actual amplification as a function of ε is illustrated in Fig. 3 for the cases $\chi = 0$ and $\chi = \theta_0/3$. In the former case the maximum is determined by φ_0. Clearly the width at given amplification is directly related to T_A. Thus by increasing the amplification factor the steepness of rise can be increased and the width of the pulse reduced. The apparent shape of the signal will then depend on the residual signal level (RSL).

If it is assumed that the whole surface of a neutron

star acts as the effective source, there is clearly an upper limit to the maximum amplification. This is because there is an upper bound M_c to the mass of a stable neutron star configuration, and a corresponding lower bound r_c to its radius. Meltzer and Thorne[3] give values of M_c and r_c for two assumed equations of state, one due to Harrison, Wakano and Wheeler (HWW), the other to Skryme, Cameron and Saakyan (SCS). Of these the latter give the most favourable values of r_c and M_c, from which it follows that the maximum possible amplification is

$$A^*_{max} \approx 17\,P^{1/3} \qquad (10)$$

The HWW values give the coefficient of $P^{1/3}$ as ~ 10. These limits can be relaxed, however, if the stars have significant angular momentum or if a further hard core pressure (the C-field[4], for instance) exists. More important, if the radiation comes from smaller regions on the stellar surface, from flares for example, the value of A^*_{max} can be exceeded by large factors. This question will be discussed further in the following section in which we attempt a comparison of our proposal with the pulsating star hypothesis.

Comparison of Binary and Oscillation Theories

We now list some of the observational and theoretical problems raised by the new radio sources and examine them in the light of the theoretical explanations so far proposed.

(i) Random Signal Strengths

It appears that the signal amplitudes vary considerably from pulse to pulse, disappearing below RSL for a considerable proportion of the time. On the binary star theory there is clearly no reason to expect correlations between successive pulse intensities because the "clock mechanism" (the gravitational orbit) is completely divorced from the energy source. Random flares occurring on each star or in the region between stars happen to be amplified, and thus observed, only at moments of suitable alignment. We note, however, that if the flare lifetimes were significantly longer than the orbital period one would expect to observe their effects over several pulses. This may be the explanation for the observation[1] that, typically, the pulses are only present for about 1 min, which may occur quite randomly within the 4 min interval permitted by the reception pattern. Indeed any evidence for (even temporary) correlations between alternate pulse strengths would argue strongly for the binary theory, since this would have a natural explanation in terms of unusual

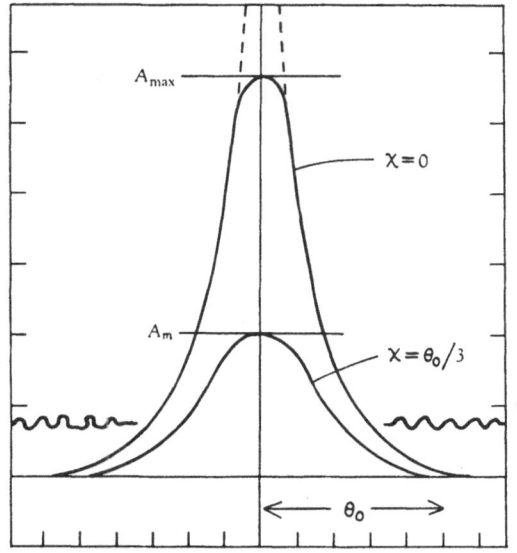

Fig. 3. Typical individual pulse shapes for angular impact parameters $\chi = 0$, and $\chi = \frac{1}{3}\theta_0$.

activity on one star only. Hewish[1] has reported that several alternate pulses were very weak for one of the objects.

On the other hand, it is difficult to understand how an oscillating star can be tuned so finely that its period changes systematically by less than ~1 part in 10^{13} per cycle, yet can trigger pulses of completely random intensity. As a general rule, the shorter and more regular the period of a vibrating star, the more precisely repetitive is the energy output as a function of phase.

(ii) Sharpness of Signal

In the binary theory the apparent sharpness of the signal peak is determined by three main factors—the degree of alignment, the ratio of pulse strength to RSL and the angular diameter of the source (see Fig. 3). As we have noted, there is a maximum possible amplification factor A^*_{max} (equation (10)) if the emissivity is uniform over the surface of the source. If, as is more likely, flares occur over smaller regions of size $\sim r_f \ll r_o$, the value of A_{max} is enhanced by a factor r_o/r_f and the energy requirements at the source are reduced by r_f/r_o. The shape or changing position of the flare on the surface of the source could lead to some signal asymmetry, though in general we would expect this to be small. The duration of the flatter portion at the peak relative to the steepness of rise and fall depends on A_{max} and the residual signal-level and could, provided the source were single, be used in the determination of these parameters.

We should, however, point out that if there were a number of flare zones on a star the signal received would consist of a quick succession of bursts of the type illustrated in Fig. 3. The duration T_c of this combined signal would be determined essentially by the time for the stellar disk to cross the cone of inversion. Thus

$$T_c \approx (1 + \varphi_0/\theta_0)\, T_1 \qquad (11)$$

If $\varphi_0 > \theta_0$, T_c could be rather longer than the times computed in the previous section. In this case, the general emission from the star would be unamplified whereas that from each flare zone would be amplified by A as given in equations (2) and (3) with r_o replaced by r_f. In this case, the sum of individual pulse lifetimes must be rather less than the time, T_c, over which the pulses are received, if the basic source is to remain below the level of detection (assumed the same for pulsing and constant sources). (In the previous section we excluded the possibility that the source is a compact white dwarf. If, however, the signals come from a small region on such a star, substantial values of A could still be obtained although the general emission would remain constant. The situation would, however, be far more complicated and is not covered by the small angle analysis.)

The oscillation theory has to explain both sharp rises and falls in the emission at the source. It is possible to imagine a situation in which shocks generated by the pulsations decay in times $\lesssim 0.016$ s, though it is not clear why this should always be the case. Nor is it clear why an oscillating star, tuned to 1 part in 10^{13} per cycle, should produce such shocks for only part of the time. Although such mechanisms may indeed exist, the explanation must certainly be much more complicated than that involving binary stars.

(iii) Probability of Observation

It is obviously necessary in both theories to estimate the probability of observing the required configurations. We take as common ground to both theories that neutron stars are the end point of evolution of stars of mass $\gtrsim 1.5\, M_\odot$. We list below the further assumptions that underlie our calculation.

(i) The luminosity function of the galaxy has remained essentially constant.

(ii) On the upper main sequence $L \propto M^3$, L being the luminosity of a star.

(iii) A star of $\sim 1\, M_\odot$ evolves completely in $\sim 10^{10}$ yr, the age of the galaxy also being of this order.

(iv) The lifetimes of radio bright neutron stars are long.

(v) a, A proportion q_1 of upper main sequence (UMS) stars become binary neutron star systems with periods in the region of a few seconds. b, A proportion p_1 of UMS stars become single neutron stars with negligible rotation.

(vi) a, Orbital planes of binary systems are randomly oriented in space. b, A proportion, p_2, of neutron stars have central densities in the correct range to give pulsation periods in the region of a few seconds.

Assumptions (i)–(iii) and (vi)a are reasonable and require no comment. We will give the remaining assumptions further attention.

From assumptions (i)–(iii), and using Allen's[5] tables, we find that n, the number density of completely evolved massive stars, is given by

$$n \approx 4 \times 10^{-3}\, pc^{-3} \qquad (12)$$

a value common to both theories.

From assumption (v) a the number density of binary neutron star pairs is nq_1. Given (vi) a, the probability of the observer lying sufficiently close to the orbital plane to see an amplification A is given by the appropriate δ_A derived from equation (7). Thus the number density n_b of suitably oriented pairs is given by

$$n_b \simeq nq_1\, \delta_A \qquad (13)$$

If $A \approx 3$, then $\delta \approx 10^{-2}$, and it follows that the expected number of favourable configurations within ~ 65 pc (the upper limit to distance given by Hewish et al.) is $\sim 40\, q_1$. We suggest that q_1 may be of order unity. In support of this, we note that the progenitors in general possess considerable angular momentum, and on contraction are likely to break up into two orbiting masses, according to a well known theory of double star formation[6,7]. Furthermore, a main sequence star is likely to reach the neutron star state via some sort of white dwarf configuration (non-static) which should be in a state of extremely rapid rotation and should break up as soon as any major contraction occurs. Thus an orbital separation of approximately a white dwarf radius ($\approx 10^8$–10^9 cm) and a period of order ≈ 1 s is not implausible. Angular momentum requirements are easily met because typically we require only $\sim 1/100$ of the angular momentum to be retained by $\sim 1/5$ of the mass. Furthermore, we note that because of the rapid rotation of UMS stars it is entirely plausible that multiple fission occurs leading to more than one neutron star binary pair[8]. We emphasize that this is extremely speculative but note however that among UMS stars the number of binaries, presumably formed during the Hayashi contraction phase, is comparable with the number of single stars. Thus a value of $q_1 \approx \frac{1}{2}$ seems not unlikely. It leads to about 20 binary neutron star systems within 65 pc having orbits sufficiently aligned for reasonable amplification. The suggested mechanism cannot therefore be excluded on probability grounds, subject, of course, to the condition that the binary system lifetime is long.

In the oscillation theory the number density of candidates n_p, corresponding to n_b is given by

$$n_p = np_1 p_2 \qquad (14)$$

The angular momentum difficulty suggests $p_1 \ll 1$. The calculations of Meltzer and Thorne[3] indicate that the pulsation period of a neutron star is extremely sensitive to the central density. Thus unless there is some particular mechanism which selects periods of 1 s it follows that $p_2 \ll 1$. (For white dwarfs it is necessary for the oscillation to be in an overtone mode if the observed period of ~ 1 s is to be obtained; this again indicates a low probability.) It therefore appears that n_p cannot significantly exceed n_b and may in fact be rather smaller. Neither theory can, however, be excluded, subject to the assumption of long lifetimes. We now consider this question.

(iv) Constancy of the Period and Energy Requirements

Hewish *et al.* have estimated the energy output of the first source to be $\sim 10^{47}$ ergs yr^{-1} provided it is at a distance of 65 pc. This is an extremely tentative figure, even apart from the uncertainty in the distance, because the spectral distribution is still not known in any detail. We will assume in the following discussion, *faute de mieux*, that this figure is correct. We defer discussion of energy radiated by gravitational waves until later in the paper.

It is important for theoretical purposes to place restrictions on the permissible amount of *systematic* change of period per cycle. From the present data it appears that the period is constant to at least one part in 10^{13} per cycle.

We remark that there is no variable red-shift problem for the binary model. The amplification can occur only when the motion is transverse. We further note that a change in mass would alter the periods in either model. The estimated changes due to accretion are, however, many orders of magnitude too small to have significant effects in either case.

If we assume that the oscillating star contains the maximum possible vibrational energy ($\sim 10^{51}$ ergs), the estimated radiation rate at 65 pc ($\sim 10^{47}$ ergs yr^{-1}) yields a decrease in vibrational energy of ~ 1 part in 10^4 yr^{-1} and presumably a similar change in amplitude. Although in a linear theory the period is independent of the amplitude of pulsations, a residual dependence on terms of higher order would be expected, particularly near maximum amplitude. The observed constancy to 1 part in 10^6 yr^{-1} appears to show a remarkable and, we think, unlikely absence of such non-linear effects. This argument can be countered by reducing the distance to the source by a factor of at least 10, but only at the expense of a further reduction in the expected number of stars to $n_p \simeq 4 p_1 p_2$. This is now disturbingly small.

The figures given here point clearly to a further difficulty. If the estimated emission rate is correct, the lifetime T_p of the pulsations can be at most $\sim 10^4$ yr. (Indeed, T_p is probably less than this because, according to Meltzer and Thorne[3], much of the available energy will be dissipated rapidly (in ~ 10 yr) by various processes including neutrino emission.) On this basis, the probable number of vibrating neutron stars of the type observed should be reduced by a further factor T_p/T_* where T_* is the mean total evolutionary time of neutron star candidates. With $T_* \sim 10^9$ yr, this would again reduce the number of possible candidates within 65 pc to unacceptably low levels. Similar arguments apply in the case of white dwarfs with corrections of order unity for the maximum possible energy and the expected number of stars. One is therefore in some difficulty to explain the observed number of sources of this type on the pulsating star hypothesis.

On the other hand, we run into very similar, though slightly less stringent, difficulties on the binary star hypothesis. Again assuming the source to be at 65 pc, our total energy requirements are increased by the ratio of orbital period to pulse duration and reduced by the amplification factor and, if the emission is coming from n_f smaller flare zones, by the factor $(r_f/n_f r_0)$. (If there were a correlation between flare activity and the position of the binary companion a further reduction would be possible.) Thus our energy requirements are probably of the same order of magnitude as those in the pulsating star theory, with the possibility of some reduction from the $(r_f/n_f r_0)$ factor. The expected number of pairs within 65 pc is thus likewise reduced to unacceptable levels.

The problem of energy supply and reasonable lifetimes thus seems insuperable unless the total emission rate at 65 pc turns out to be a considerable over-estimate. Otherwise only a harder core pressure than that from neutron degeneracy seems to offer any hope of resolving this difficulty. A possibility arises from the work by Hoyle and Narlikar[4] on the resistance to compression offered by the C-field. In this case the masses (to which

there is no upper bound) are near their Schwarzschild radius. Clearly this topic requires a more detailed examination which is beyond the scope of this article. We merely note that to maintain the number density the mass cannot exceed $\sim 3\ M_\odot$ and the energy reserve not more than 10^{55} ergs. Thus even at the extreme, the lifetimes are only just comparable with T_* and the problem of expected probable numbers is barely resolved. We therefore feel that if either of these pictures is to prove correct the total emission rate at 65 pc must be reduced by at least $\sim T_*/T_p \simeq 10^5$.

Whatever the figure for total emission rate turns out to be, we would, at this point, emphasize one important advantage of the binary star picture over the pulsating star hypothesis. In the latter, any energy taken from internal reserves causes a change in structure and hence in period on the time scale T_p. Thus even if the energy requirement is reduced so that $T_p \sim T_*$ the periods will cover a wide range, ensuring a small value for p_2 and hence of the expected number of sources. On the binary picture, however, these internal reserves and indeed energy stored in the region between the stars (in magnetic fields, for instance) could be utilized without altering the periods, provided the rest mass of the stars remains the same. Even at the estimated emission of $\sim 10^{47}$ ergs yr^{-1} this gives a maximum change in orbital period of 1 part in 10^7 yr^{-1}—still within the observational limits. We have, however, omitted any consideration of gravitational radiation throughout the analysis. We now examine this question in more detail.

Gravitational Radiation

There is still some controversy as to the precise situations in which gravitational waves are emitted. We follow the standard analysis of Landau and Lifshitz[9], who obtain for the characteristic time τ_B for changes in the binary orbit the expression

$$\tau_B \approx \frac{1}{128} \frac{c^5 l_{OD}{}^4}{G^3 M_1 M_2 (M_1 + M_2)} = \frac{1}{128} \frac{c^5 (M_1 + M_2)^{1/3} \omega^{-8/3}}{G^{5/3} M_1 M_2} \quad (15)$$

Using the approximate values $M_1 = M_2 = 0.5\ M_\odot$ and $l_{OD} \sim 3 \times 10^8$ cm as before, we obtain $\tau_B \approx 10^8$ s. By reducing one of the masses somewhat τ_B could be increased to at most $\sim 10^9$ s. This clearly implies a relative change of period far in excess of the observational limits and seems to exclude the binary star theory quite conclusively. Before proceeding, however, we would point out that the pulsating neutron star hypothesis also suffers in such an analysis.

For a pulsating star the characteristic decay time due to gravitational radiation is given[9,10] by

$$\tau_p \simeq \frac{50\ c^5}{G M a^2 e^2\ \omega^4} \quad (16)$$

where e is the ellipticity. As the loss rate is dominated by the highest harmonics present (16) could be an overestimate. For a neutron star with a period of ~ 1 s we obtain[3] $M \sim 0.2\ M_\odot$, $a \sim 2 \times 10^7$ cm, and thus $\tau_p \simeq 2 \times 10^9/e^2$ s. Should neutron stars retain only ~ 1 per cent of the angular momentum of typical UMS stars, they would have marked ellipticity, $e \sim 0.1$, resulting in a decay time τ_p of at most $\sim 10^4$ years. (Similarly a white dwarf with solar angular momentum would have $\tau_p \sim 10^5$ years). Such time scales are already incompatible with the observational data. There is, however, a suggestion due to Tsuruta and Cameron[11] that a strong magnetic field could remove sufficient angular momentum to reduce e to a negligible value. From their analysis, however, the required magnetic field would itself produce a substantial quadrupole moment. Thus, unless a fine balance exists in which first angular momentum and then magnetic fields are dissipated rapidly, the pulsating neutron star would lose its energy on an extremely short time-scale. The number of stars on the upper main sequence with low (effectively zero) angular momentum could produce the required type of neutron star but only with immensely reduced probability of observation. We feel that if the theory used here is correct the pulsating neutron star hypothesis can be excluded. Pulsating white dwarfs would seem to have the least difficulty concerning gravitational radiation, the associated time-scales always being $\gtrsim \tau_B$. Even in this case, however, one would expect changes in period on a time-scale of $\sim 10^{12}$ s (because most of these stars should be in rapid rotation even if on the main sequence they possessed only the solar angular momentum). This should also have been detected with the present observational data.

Both the binary star and the pulsating star models seem to succumb under the problem of gravitational radiation. On the other hand, our model appears to satisfy a number of other observational requirements in a rather natural way. If observations continue to support the binary star picture, and if general relativity is fundamentally correct, there seem to be two possible ways to resolve the gravitational radiation difficulty.

First, it may be that freely falling particles (binary stars, for example) do not radiate gravitational energy according to general relativity. This possibility has been urged by Bondi[12], who found that time dependent systems without "news" (that is, radiation of energy) can exist. The physical interpretation of his solutions remains controversial but, in view of the recent observations of rapidly pulsing radio sources and their otherwise plausible interpretation as binary stars, it is important that Bondi's suggestion be investigated in detail. We note, however, that this would not resolve the gravitational radiation problem in the case of pulsating neutron stars because the mass in these bodies is not in free fall.

The second possibility is that quantization effects become important in fairly high fields. The gravitational radiation difficulty is similar to that which faced spectroscopy before the Bohr theory. While physicists have long recognized the desirability of quantizing general relativity, no generally acceptable scheme has been proposed. To some extent, this failure may be due to lack of experimental or observational guidance for strong field situations. If our interpretation of these radio sources is correct, the remarkable constancy of the orbital period leads one to suspect some sort of quantization mechanism; that is, radiation free orbits, possibly arising as eigensolutions in the strong field (non-linear) equations. It may be significant that $v/c \approx 1/100 \approx \alpha$ where α is the electromagnetic fine structure constant. Could the observations be hinting that the usual strong field quantization condition $GM/rc^2 \approx 1$ is unrealistic and that $GM/rc^2 \approx \alpha$ is more relevant? We cannot pursue the matter further here, but it seems possible that once again astronomy may be able to provide a laboratory for testing consequences of the general theory of relativity, but this time in strong fields.

Predictions of the Binary Theory

The detailed consequences of our suggestion are now being evaluated. At the present stage, we are able to make some general predictions (applicable to neutron star pairs) which may act as further tests of the proposal.

(a) *For a single pulse arising from a homogeneous source region.* (i) The times of onset and disappearance should be the same at each frequency for measurements between given amplification levels. (ii) The duration of a pulse $\lesssim 2 \times 10^{-2} T$, where T is the time interval between repeated pulses (that is, $\sim 1 \cdot 3$ s in the present case). (iii) The *amplification* should be the same at all frequencies. It should be noted that in applying (i)–(iii) due allowance must be made for the residual signal level, which at some frequencies may swamp the amplified signal and cause spurious variations in the predicted temporal coincidences.

(b) *For multiple pulses.* In general, the original, unamplified signal may have two different components coming from (i) the whole surface and (ii) flare activity in concentrated regions. Both could be time varying. They should produce an observed signal as follows: (i) A general rise in the entire signal by a modest amount $\lesssim 17$ (cf. section 3) lasting at the lowest amplification levels for $\sim 2 \times 10^{-2} T$ s. (ii) Sharply peaked signals with maximum amplification factors which may be $\gtrsim 17$ and with widths correspondingly narrower.

(c) The integrated unamplified signal received in time T should be comparable to the energy received in each pulse: thus the basic source should be detectable. However, we understand, that this test is quite difficult because of confusion with other, non-pulsing, sources in the area. One would expect to have similar difficulties in detecting the remaining $\sim 4 \times 10^3$ pairs within 65 pc. which have alignments unsuitable for amplification.

(d) If a source can be identified optically, then its light variations should either show sharp bursts similar to those observed at radio frequencies, or no amplification at all. It should not show smooth light variations as one might expect from a pulsating star.

(e) If the radio signals are short compared with the interval between them, as in the present case, an optical spectrum should not show classical white dwarf features. Although neutron star spectra are at present unknown one might expect them to exhibit fairly featureless optical continua. Should this be found, it might prove worthwhile to reinvestigate the nature of objects with such spectra, hitherto supposed to be a sub-set of the white dwarfs.

We thank Professor F. Hoyle, Dr D. W. Sciama, Professor Martin Ryle and Dr A. Hewish for helpful discussions. One of us (W. C. S.) carried out this work during tenure of an NSF Pre-doctoral Fellowship and a Jesus College, Cambridge, research fellowship. Two authors (J. F. and P. A. S.) were in receipt of William Stone research fellowships at Peterhouse, Cambridge.

Note added in proof. Since this article was prepared, further observations have been reported by Ryle and Bailey[13] and Davies, Horton, Lyne, Rickett and Smith[14]. Ryle and Bailey's provisional identification of the source with a blue star-like image of $\sim 18^m$ clearly offers the possibility of testing predictions (d) and (e) above. The results of Davies et al., in particular the simultaneous onset of emission at all observed frequencies, and the appearance of multiple-pulsing with three or more single pulses per complete pulse, strongly support the binary view put forward in this paper. Further observations of individual pulse shapes at several frequencies could distinguish between a pulsating theory in which plasma oscillations (Ryle and Bailey) produce radio bursts and the binary theory. Frequency dependent collisional damping with a time constant αv^{-2} could produce substantially different pulse widths at different frequencies. A v^{-2} width-dependence would not be so easily explained on the binary hypothesis.

Received March 11, 1968.

[1] Hewish, A., Bell, S. J., Pilkington, J. D. H., Scott, P. F., and Collins, R. A., *Nature*, 217, 709 (1968), (Paper 1); and Hewish, A., Cavendish Colloquium, 20 February, 1968.

[2] Liebes, jun., S., *Phys. Rev.*, 113 B, 835 (1963).

[3] Meltzer, D., and Thorne, K. S., *Ap. J.*, 145, 514 (1966).

[4] Hoyle, F., and Narlikar, J. V., *Proc. Roy. Soc.*, 278, 465 (1963).

[5] Allen, C. W., *Astrophysical Quantities*, second ed., 238 (Athlone Press, London, 1963).

[6] Jeans, J., *Astronomy and Cosmogony*, second ed. (Dover Publ., 1961).

[7] Lyttleton, R. A., *The Stability of Rotating Liquid Masses* (Cambridge Univ. Press, 1953).

[8] Hoyle, F., Narlikar, J. V., and Wheeler, J. A., *Nature*, 203, 914 (1964).

[9] Landau, L. D., and Lifshitz, E. M., *Classical Theory of Fields* (Addison Wesley, 1951).

[10] Zee, A., and Wheeler, J. A., as in *Ann. Rev. Astro. Astrophys.*, 4, 393 (1966).

[11] Tsuruta, S., and Cameron, A. G. W., *Nature*, 211, 356 (1966).

[12] Bondi, H., *Proc. Roy. Soc.*, A, 269, 21 (1962).

[13] Ryle, M., and Bailey, J. A., *Nature*, 217, 907 (1968), (Paper 19).

[14] Davies, J. G., Horton, P. W., Lyne, A. G., Rickett, B. J., and Smith, F. G., *Nature*, 217, 910 (1968), (Paper 6).

43. Pulsed Radio Sources

by

G. R. BURBIDGE
University of California at San Diego,
La Jolla, California

P. A. STRITTMATTER
Institute for Theoretical Astronomy,
University of Cambridge

RADIO astronomers at the University of Cambridge have recently discovered a class of pulsed radio sources with the following properties: (i) a remarkably constant period of ~ 1 s; (ii) a pulse duration of ~ 20–40 ms; and (iii) an apparently random pulse intensity[1,2]. Further investigation has shown that the main pulses may have an internal structure of sub-pulses of ~ 10 ms duration[3,4] and that the radiation, at least in a single sub-pulse, may be highly polarized[5].

At this early stage, there are clearly many possible interpretations of the phenomena. Most attention has so far been concentrated on providing a clock mechanism with the observed period, which immediately seems to demand extremely condensed systems. Thus current proposals involve pulsating white dwarfs[6], rotating white dwarfs[7] and neutron star binary systems[8]. None of these can yet be excluded though the first two are already being pressed to the lower limit of possible periods while the third may encounter difficulties associated with gravitational radiation. None of the above proposals naturally includes an explanation of the pulse substructure or the polarization, however, though there is no reason why these should not be produced in some mechanism extraneous to the clock. The gravitational lens binary model[8] seems to have most freedom in this respect because the energy production and clock mechanisms are in any case separated. In this note we wish to suggest an alternative model for producing the pulsed radiation which would include pulse structure and polarization in a natural way.

It has been known for almost two decades that the planet Jupiter emits bursts of radiation at decametric wavelengths and that these occurrences are correlated with certain orientations of Jupiter or, more particularly, its magnetic field, with respect to the Earth–Jupiter line[9]. More recently, Bigg[10] and Dulk[11] have shown that the emission is also correlated with the position of the satellite Io in the sense that strong bursts are virtually certain when both Io and the magnetic field axis are simultaneously in favourable positions. While there have been a number of attempts to explain these observations, in terms of a plasma wake or distortions of Jupiter's magnetic field by the presence of the satellite, it seems that a generally acceptable and comprehensive theory has yet to be proposed. Indeed, even the precise location of the emitting regions is still in some doubt. None the less, the very existence of such a pulsing mechanism suggests that it may well prove profitable to look for an analogous interpretation for the pulsars.

We therefore propose that the central object is a neutron star of mass $M_c \sim 0.5$ $M_\odot$ and that it possesses a small satellite with an orbital period P_0 of ~ 1 s, so that it has an orbital separation of $\sim 10^8$ cm. This satellite may be formed from remnants of the supernova explosion which presumably took place when the neutron star was formed; this would then account naturally for the ~ 1 s period[8,12]. Provided the mass M_s of the satellite $\leqslant 10^{27}$ g, no difficulties need arise from radiation of energy by gravitational waves. Further, the mass of the satellite does not seem to be crucial so far as the disturbance to the magnetosphere is concerned because even artificial earth satellites produce measurable effects[13].

The central star, which represents the final evolved state of an upper main sequence star, should be in a state of rapid rotation with a period P_r somewhat in excess of 1 ms depending on the mass. The main pulse may thus be envisaged as arising from the favourable location of the satellite (period P_0) while the sub-pulses are to be associated with the passage of a magnetic field axis through the satellite (period P_r). The observations of CP 1919, for example, would suggest a value $P_r \sim 20$ ms. (If, however, we assume that the central star rotates near its break-up velocity, its mass may be determined from P_r and, hence, the satellite orbital distance from P_0.)

Clearly it is necessary that the emitted radiation be beamed, the required cone for the pulsars having a semi-apex angle of $\sim 5°$, rather smaller than in the case of Jupiter. This causes no difficulty concerning probability of observation[8]. The question arises, however, as to whether there are one or more positions on the orbit from which radiation can be received. For simplicity we assume there to be one only, namely when the satellite is moving more or less towards the observer. In this context we note, following Gold, that the velocity of plasma corotating with the central star would be approaching that of light at the orbit of the satellite. At this stage the situation becomes most complex. We note, however, that these high velocities would necessarily result in preferential emission in the forward direction, thus assisting any beaming mechanisms of the Jupiter type which may already be operative. Also the local magnetic field would be perpendicular to the line of sight at the point of emission resulting possibly in linearly polarized light. The direction of polarization would "rock" as in the case of Jupiter's decimetric radiation if the field were inclined to the rotation axis as assumed implicitly above.

A difficulty arises in estimating the magnetic field. The sharp drop-off in energy output at the higher frequencies[14] suggests that the local cyclotron frequency ν_c must satisfy $\nu_c \sim 100$ MHz (or less if relativistic beaming is important); thus the local magnetic field strength $H_0 \sim 30$ gauss. Assuming a typical neutron star radius ~ 10 km and a dipole field, it follows that the field H_s at the stellar surface $\sim 3 \times 10^7$ gauss. (A multipole field would, of course, allow much higher values of H_s.) This compares with a value $\sim 10^9$ gauss from direct compression of a "general" solar type magnetic field and the figures of $\sim 10^{14}$–10^{16} gauss suggested by Woltjer[15] for the maximum allowable field strength in a neutron star. On the other hand, it is by no means clear that a magnetic field can continue to exist in highly degenerate material. The largest field that can be confined in the outer non-degenerate regions is given by

$$B^2/8\pi l \approx GM\bar{\rho}/r^2$$

where l and $\bar{\rho}$ are the thickness and mean density of these zones; this again gives $H_s \sim 10^9$–10^{10} gauss. There is also the question of whether this field can maintain itself against decay or whether a neutron star field would need to be dynamo maintained by the rapid rotation. Clearly these questions require closer analysis. We note, however, that with the observational requirements on electron density and magnetic field in the region of the satellite orbit magnetic forces could still balance inertial forces, and there would be no significant loss of angular momentum from the star.

The radiating particles in this model would have to be supplied internally rather than through an external agency (the solar wind) as in the case of Jupiter. Thus the energetic electrons could either be generated through the interaction of the satellite with the corotating plasma (a satellite of radius $\gtrsim 10^6$ cm could just produce the observed radio emission $\sim 10^{26}$–10^{27} erg s^{-1} without demanding too large a plasma density or running into momentum difficulties) or, perhaps more likely, be supplied from the surface of the star during field distortions induced by the satellite. The intensity of the pulses would clearly depend on the supply rate of particles which would in turn depend on flare activity, etc., on the central star. Should there be a substantial number of trapped particles there should also be continuous emission at higher frequencies, analogous to Jupiter's decametric radiation. However, using the Jupiter analogy once more, the flux would probably be well below present detection levels. Finally, we note that the precise onset of an individual pulse would be determined largely by the beaming mechanism but to a lesser extent by the rotation of the central star. Some variation of pulse onset time

($\sim$ a few ms) might therefore be expected from pulse to pulse, though no systematic effect should be present.

While we recognize that the present "theory" of pulsars is sketchy in the extreme, we note that it would allow a qualitative understanding of the pulse period, the subpulse duration, the polarization and possibly to some extent the beaming of radiation. Unfortunately the details of all but the first of these depend sensitively on the magnetic field configuration which we are unable to predict (but which the observations should in principle allow us to determine). Although progress along these lines will be difficult, the similarity between this and another well observed but ill-understood phenomenon, the radio bursts from Jupiter, is encouraging.

We wish to thank Prof. F. Hoyle and Drs J. Faulkner and J. V. Narlikar for helpful discussions.

Received April 23, 1968.

[1] Hewish, A., Bell, S. J., Pilkington, J. D. H., Scott, P. F., and Collins, R. A., *Nature*, 217, 709 (1968), (Paper 1).
[2] Pilkington, J. D. H., Hewish, A., Bell, S. J., and Cole, T. W., *Nature*, 218, 126 (1968), (Paper 2).
[3] Davies, J. G., Horton P. W., Lyne, A. G., Rickett, B. J., and Smith, F. G., *Nature*, 217, 910 (1968), (Paper 6).
[4] Drake, F. D., *Science* (in the press).
[5] Lyne, A. G., and Smith, F. G., *Nature*, 218, 124 (1968), (Paper 7).
[6] Thorne, K. S., and Ipser, J. R., *Ap. J. Lett.* (in the press).
[7] Ostriker, J. P., *Nature*, 217, 1227 (1968), (Paper 39).
[8] Saslaw, W. C., Faulkner, J., and Strittmatter, P. A., *Nature*, 217, 1222 (1968), (Paper 42).
[9] Warwick, J. W., *Ann. Rev. Astron. Astrophys.*, 2, 1 (1964).
[10] Bigg, E. K., *Nature*, 203, 1008 (1964).
[11] Dulk, G. A., *Science*, 148, 1585 (1965).
[12] Hoyle, F., Narlikar, J. V., and Wheeler, J. A., *Nature*, 203, 914 (1964).
[13] Tiuri, M. E., *Planet. Space Sci.*, 15, 1203 (1967).
[14] Moffet, A. T., and Ekers, R. D., *Nature*, 218, 227 (1968), (Paper 9).
[15] Woltjer, L., *Ap. J.*, 140, 1300 (1964).

44. Some Models for Pulsed Radio Sources

by

F. PACINI
E. E. SALPETER

Center for Radiophysics and Space Research,
Cornell University,
Ithaca, New York

ALREADY before the discovery of pulsed radio sources, the possibility had been discussed[1,2] of rotating neutron stars providing an energy source in supernova remnants. A large number of models have been proposed to explain the pulsed radio sources and some of these invoke a central neutron star, some kind of rotation and some kind of inhomogeneity. In the preceding article Gold has proposed a model in which the neutron star, surrounded only by a tenuous plasma, rotates itself with the inhomogeneity residing merely in the star's magnetic field. Another model[3] considers the possibility that the rotation and inhomogeneity both reside in the orbital motion of a satellite around a neutron star. We discuss here some difficulties associated with models for pulsed radio sources which invoke the orbiting of material satellites around any central object.

We are chiefly interested in a satellite orbiting around a neutron star, but shall discuss more generally the possibilities of a satellite of mass m orbiting around a more massive central "star" of any mass M (including the unlikely case of a neutron star orbiting around an invisible object of very large mass). The radius a of the orbit is, of course, related to the observable period T of the orbit by Kepler's laws. With masses m and M in solar units and T in seconds

$$a = 1 \cdot 5 \times 10^8 M^{1/3} T^{2/3} \text{ cm} \tag{1}$$

We first show how small the satellite mass m has to be in order to avoid noticeable changes in the orbital period due to gravitational radiation.

The total energy, E, of the Kepler system is $(-\mathrm{G}Mm/2a)$. If the only energy loss of the system is that due to the radiation of gravitational waves, the time rate of change of energy is given by

$$\frac{\mathrm{d}E}{\mathrm{d}t} = -\frac{32}{5} \frac{\mathrm{G}\mu^2}{c^5} a^4 \left(\frac{2\pi}{T}\right)^5 \tag{2}$$

where $\mu = Mm/(M+m)$ is the reduced mass. Approximating μ by m (assuming $m \ll M$) and eliminating a by means of equation (1), we find

$$\frac{1}{T_{\mathrm{rad}}} \equiv \frac{3}{2} \frac{1}{E} \frac{\mathrm{d}E}{\mathrm{d}t} \simeq 4 \times 10^{-6} \, m \, M^{2/3} \, T^{-8/3} \tag{3}$$

Because the period T is proportional to $(-E)^{-3/2}$, it decreases by a fraction of (t/T_{rad}) during a time-interval T. Changes in T have not yet been detected and the observations indicate $T_{\mathrm{rad}} > 10^{14}$ s. Masses of neutron stars are larger than $0 \cdot 03 \, M_\odot$, and so we need $m < 3 \times 10^{-8} \, M_\odot \sim 10^{26}$ g if $T \sim 1$ s.

If the central star of mass M is itself rotating and the orbital period T of the satellite is locked in with this spin, the rate of change of T is slower. We can still, however, get an upper limit to m as follows. The radius of the central star must be less than the radius a of the satellite orbit and its rotational energy less than $E(M/m)$. The rate of energy loss in gravitational radiation is the same as before and we have

$$\frac{1}{T_{rad}} > 4 \times 10^{-6} \, m^2 \, M^{-1/3} \, T^{8/3} \, s^{-1} \qquad (4)$$

The orbital radius a must be larger than the Schwarzschild radius $(2GM/c^2)$ of the central body (for stable orbits one actually needs $a > 6GM/c^2$), so that

$$M < 3 \times 10^4 \, M_\odot \left(\frac{T}{1 \, s}\right) \qquad (5)$$

We then get as an extreme limit

$$\frac{1}{T_{rad}} > 10^{-7} \, m^2 \, T^{-3} \, s^{-1} \qquad (6)$$

With $T_{rad} > 10^{14}$ s and $T \sim 1$ s we have $m < 3 \times 10^{-4} \, M_\odot$, so that the satellite could not be as massive as a neutron star, no matter what the central body is.

We have further problems on the internal structure of the satellite. If the satellite is held together purely by gravitational forces, it must be orbiting outside the Roche limit, that is, we must have $GMa^{-3} < Gmr^{-3}$, where r is the radius of the satellite. Using equation (1) we then find for the mean density of the satellite

$$\rho > (1.4 \times 10^8 \, g \, cm^{-3}) \left(\frac{1 \, s}{T}\right)^2 \qquad (7)$$

independent of the mass M of the central body. Neutron stars and the more massive white dwarfs can achieve such high densities, but no object of mass $m \ll 0.01 \, M_\odot$ can. This lower limit on m exceeds by far the upper limit from equation (6). A satellite held together by gravitational forces is therefore impossible no matter what the mass of the satellite or of the central body.

We have to consider next satellites held together by solid-state forces against the tidal effects of the central star. The tension in the centre of the satellite is proportional to ρr^2 and we can find an upper limit for the satellite radius r for it to be stable against disruption. Using equation (1) to eliminate a, we find an expression independent of M

$$r = 7.3 \, m \left(\frac{\rho}{2 \, g \, cm^{-3}}\right)^{-\frac{1}{2}} \left(\frac{S}{10^8 \, dyne \, cm^{-2}}\right)^{\frac{1}{2}} \frac{T}{1 \, s} \qquad (8)$$

where S is the tensile strength of the material. This gives an upper limit of the order of 10 m for most natural materials (only about 100 m even for good structural steel).

A single satellite of size 10 m could hardly produce a sufficient disturbance on the magnetosphere of the central star. One might then think of a collection of a large number of such small satellites, but this raises other problems: gravitational forces could hold together a conglomeration of a number of such small objects all at the same orbital radius a but not, as we have seen, those with different values of a. It is not easy to see how one could single out values of a in some very narrow range only, and one would therefore expect a spread in orbital periods T.

The requirement of tensile strength of the orbiting bodies requires them to be solid which requires their temperature to be below about 10^3 °K, which in turn puts an upper limit on the luminosity L of the central neutron star. With $M \lesssim 2M_\odot$ and $T \sim 1$ s this gives $L < 10^{-8} \, L_\odot$. An undisturbed neutron star could cool to this low optical luminosity in a few times 10^7 yr, but with a large number of small satellites orbiting the neutron star and interacting with the magnetosphere one would expect a steady stream of these objects to spiral into the star and heat it. To keep the neutron star luminosity below $10^{-8} \, L_\odot$, one would require a rate of infalling material of less than 10^{13} g/yr. To summarize: the possibility of solid objects orbiting around a central neutron star appears highly unlikely.

We thank Professor T. Gold for pointing out the limitation arising from Roche's condition and Professor G. Burbidge for communicating unpublished work. This work was supported in part by the US National Science Foundation and in part by the US Office of Naval Research.

Received May 20, 1968.

[1] Wheeler, J. A., *Ann. Rev. Astron. and Astrophys.*, 4 (Ann. Res., Inc., Palo Alto, California, 1966).

[2] Pacini, F., *Nature*, 216, 567 (1967).

[3] Burbidge, G. R., and Strittmatter, P. A., *Nature*, 218, 433 (1968), (Paper 43).

45. Pulsar Io Effect?

by

D. H. DOUGLAS-HAMILTON

Harvard College Observatory,
Cambridge, Massachusetts

Burbidge and Strittmatter[1] have suggested that a small orbiting companion of a neutron star might trigger pulses of radio emission, in a way analogous to the Io effect in Jupiter. They propose a satellite of mass $m < 10^{27}$ g, to avoid difficulties with gravitational radiation, orbiting at a distance $R = 10^8$ cm to give a period of the order observed in pulsars.

At this distance the tidal stresses impose a minimum (the Roche limit) on the satellite density ρ, given by

$$\rho > 14.8 \, (R/r)^3 \, \rho_c$$

where r and ρ_c are the radius and mean density of the central body. With $r = 10$ km, $\rho_c = 10^{14}$ g cm^{-3}, this condition gives

$$\rho > 1.4 \times 10^9 \, g \, cm^{-3}$$

Such a density implies a highly degenerate configuration. Neutron stars, however, have a lower mass limit because they can only be stable if

$$\frac{Gm_n M}{r'} > \Delta$$

where r' is the radius and M is the mass of the star, m_n is the mass of the neutron, and Δ is the difference in binding energy per nucleon between neutron and atom (say iron)[2]. With smaller mass the formation of atomic nuclei is energetically favoured. Oppenheimer and Volkoff[3] give the limit as $0.1 \, M_\odot$; Landau and Lifshitz[2] give it as $1/3 \, M_\odot$. A neutron star of mass 10^{27} g is therefore thermodynamically impossible. A white dwarf electron degenerate configuration is also inadmissible for such a low mass.

The gravitational radiation from a binary star gives a characteristic decay time for the orbit of[4]

$$\tau = c^5 \left[(m_1 + m_2)/(G^5 \omega^8) \right]^{1/3} \times 1/128 \times 1/(m_1 m_2)$$

Saslaw et al.[4] give $\tau = 10^8$ s for $m_1 = m_2 = 0.5\ M_\odot$. For $m_1 = 0.10\ M_\odot$, $m_2 = 0.5\ M_\odot$ and identical period, the characteristic time is still only

$$\tau = 4 \times 10^8\ \text{s}$$

Thus any orbiting mechanism for pulsars appears to be excluded, unless Bondi's[5] suggestion of non-radiating states in free-fall motion is accepted.

Received May 13, 1968.

[1] Burbidge, G. R., and Strittmatter, P. A., *Nature*, **218**, 433 (1968), (Paper 43).
[2] Landau, L. D., and Lifshitz, E. M., *Stat. Phys.*, 341 (1958).
[3] Oppenheimer, J. R., and Volkoff, G., *Phys. Rev.*, **55**, 374 (1939).
[4] Saslaw, W. C., Faulkner, J., and Strittmatter, P. A., *Nature*, **217**, 1222 (1968), (Paper 42).
[5] Bondi, H., *Proc. Roy. Soc.*, A, **269**, 21 (1962).

46. Amplitude Variations in Pulsed Radio Sources

by

P. A. G. SCHEUER
Mullard Radio Astronomy Observatory,
Cavendish Laboratory,
University of Cambridge

THE radio pulses from the sources discovered by Hewish et al.[1,2] show extremely regular repetition rates but quite erratic amplitude fluctuations, over time scales ranging from less than a second[1] to hours[3] or even days[1]. The rapid pulse-to-pulse variations appear to be well correlated on different frequencies[3], and, as will be shown, they are most probably an intrinsic property of the source. The long-period variations show little, if any, correlation between different frequencies[3], and it seems difficult to imagine a source of wide-band pulses which have a very complex spectrum that is preserved over hundreds or thousands of pulses. Such complex frequency structure rather suggests that the emission process is modulated by interference phenomena, either between a number of coherent pulsed sources on the surface of a star, or else between a number of images of the source caused by irregular refraction in an interstellar medium. The latter possibility is considered in this article.

For detailed discussion of scattering in irregular media the reader is referred to Uscinski[4] and for the correlation of scintillations at different wavelengths to Little[5], and to references quoted in these. Only rough calculations will be attempted here.

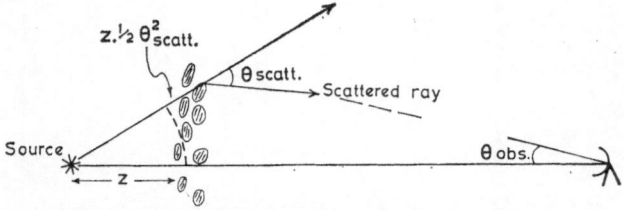

Fig. 1.   Geometry of source, screen and observer.

The interstellar medium contains large regions of ionized gas with $10^{-1\pm1}$ electrons cm⁻³; the clearest evidence for the existence of such a general distribution of free electrons comes from the rotation measures of polarized radio sources and from the dispersion of the pulses from the pulsed radio sources themselves[1,3]. Suppose the medium between us and the source is not perfectly regular, but has irregularities characterized by an electron density perturbation ΔN and a linear scale a; the phase perturbation on passing through one such irregularity is

$$\delta\varphi = \Delta N r_0 \lambda a \qquad (1)$$

at wavelength λ, where r_0 is the classical electron radius

(2.8×10^{-13} cm). On passing through a depth L of such irregularities the root mean square phase deviation becomes

$$\Delta\varphi \simeq (L/a)^{\frac{1}{2}} \delta\varphi \simeq (La)^{\frac{1}{2}} \Delta N r_0 \lambda \qquad (2)$$

Because the observed amplitude variations are large, any scattering screen which plays an important part must have $\Delta\varphi > \pi$ radians, so that nearly all the incident radiation is scattered at least once.

For most purposes the scattering is equivalent to that of a physically thin screen, causing phase deviations $\Delta\varphi$ given by equation (2), still on a linear scale a, and placed somewhere near the middle of the depth L of the actual screen[6]. Because $a \gg \lambda$ and we require $\Delta\varphi > \pi$, geometrical optics may be used to estimate the effect of the screen; incident rays are refracted through angles of the order of

$$\theta_{\text{scat}} \simeq (\lambda/2\pi)\Delta\varphi/a = (1/2\pi)(L/a)^{\frac{1}{2}}\Delta N r_0 \lambda^2 \qquad (3)$$

(If the density perturbation in each irregularity has a Gaussian spatial variation $\exp(-r^2/a^2)$, and θ_{scat} and ΔN are interpreted as root mean square values, the correct result is

$$\theta_{\text{scat}} = 2^{-1/4}/\pi^{-3/4}(L/a)^{\frac{1}{2}}\Delta N r_0 \lambda^2 \qquad (3a)$$

Note that if only a fraction f of space in the scattering layer is occupied by irregularities, the only effect is to replace ΔN by $\Delta N \sqrt{f}$ throughout our formulae.)

If the screen is placed at a distance z from the source, where $z \ll$ (distance of source from Earth), scintillations will be observed subject to the following conditions:

(i) $\Delta\varphi > \pi$, as noted earlier.

(ii) Amplitude variations (as opposed to mere refraction) will be observed only if several interfering beams reach the observer, that is, only if the region of the screen contribution to the observed radiation is larger than the size of one irregularity

$$z\theta_{\text{scat}} > a$$

(iii) The scintillations will be similar at wavelengths λ and $\lambda + \Delta\lambda$ if the phase differences between the various interfering beams are the same at $\lambda + \Delta\lambda$ as at λ, to within π radians, say. The phase differences are made up of two parts; the first, caused by beams passing through different parts of the screen, is about $\Delta\varphi$; the second, caused by different physical path lengths, is of the order of $(2\pi/\lambda)(\frac{1}{2}\theta^2_{\text{scat}}z)$. From equation (3) and condition (ii) it is easily seen that the latter contribution is greater. Thus the scintillations are correlated over a bandwidth

$\Delta\lambda$ given by

$$\Delta\lambda/\lambda = \lambda/(z\theta^2_{scat}) \qquad (4)$$

(The estimate (4) corresponds to making Little's[5] parameter $K = \pi$, so that his "bandwidth visibility" is about 0·5.) This condition is much stronger than a similar one, that the rise time of the pulses must not be drawn out more than a few ms by arriving by different paths.

Formally identical restrictions may be derived by similar arguments when the screen is at a distance z from the observer and the source is much further away.

The conditions (i) to (iii) show what ranges of ΔN and a could produce large amplitude variations coherent over a bandwidth $\Delta\lambda$. They may be represented uniquely in terms of the dimensionless scaled quantities

$$\Delta N^* = \Delta N L^{\frac{1}{2}} r_0 \lambda^{5/4} z^{1/4} \qquad (5)$$
$$a^* = a z^{-\frac{1}{4}} \lambda^{-\frac{1}{2}}$$

when they become

(i) $\Delta N^* a^{*\frac{1}{2}} > \pi$; (ii) $\Delta N^*/a^{*3/2} > (2\pi^3)^{1/4}$;

(iii) $\Delta\lambda/\lambda = (2\pi^3)^{\frac{1}{2}} a^*/(\Delta N^*)^2$ (6)

as shown in Fig. 2.

Scintillation by the Interstellar Medium

In the case of the general interstellar medium $z \simeq L \simeq$ (distance of source), and Fig. 3 shows the conditions (i), (ii) and (iii) for this case, taking the distance to be $1·5 \times 10^{23}$ cm and $\lambda = 370$ cm. It seems that even very small fluctuations in an assumed mean density of 0·1 electrons cm^{-3} can cause significant scintillation.

Observations of other radio sources already place severe limits on the irregularities in the interstellar gas; the strongest of these is the observation that the small diameter source in the Crab nebula has an apparent diameter less than 0·4″ arc at 38 MHz, so that $\theta_{scat} < 2 \times 10^{-6}$ radians for $L = 3 \times 10^{21}$ cm, $\lambda = 790$ cm, and therefore $\theta_{scat} < 10^{-7}$ radians for $L = z = 1·5 \times 10^{20}$ cm, $\lambda = 370$ cm. This restriction is also shown in Fig. 3; it adds nothing to the condition that scintillations correlate over at least 0·5 MHz near 81·5 MHz.

Time Scale of the Scintillations

Two further restrictions must now be imposed: the scintillations must not be smoothed out by the finite diameter of the source, and the time scale τ of the scintillations must correspond with the time scale of some observed amplitude fluctuations.

The irregular screen produces a diffraction pattern in space; fluctuations are observed only because either the Earth moves through the diffraction pattern, or because relative motion of the source and the screen sweeps the pattern past us. It will be assumed that the relevant

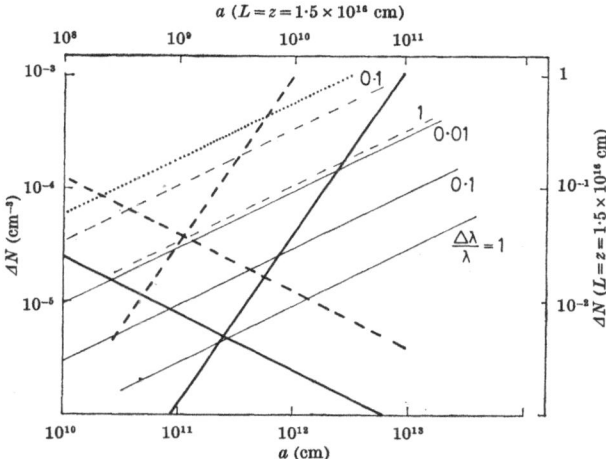

Fig. 3. Conditions required for large amplitude fluctuations in a source observed through $1·5 \times 10^{20}$ cm of interstellar medium, shown as in Fig. 2. Full lines: 81·5 MHz; dashed lines: 408 MHz. The dotted line shows the upper limit on the strength of interstellar irregularities obtained from observations of the Crab Nebula. If only a region of radius $1·5 \times 10^{16}$ cm around the source causes the scintillations, the conditions for the medium are given by the same diagram when used with the top and right-hand scales for ΔN and a.

velocities v do not greatly exceed 30 km s^{-1} in either case.

First consider the Earth's motion.

$$v\tau = \text{scale of diffraction pattern} = \lambda/\theta_{obs}$$

where θ_{obs} is the angular spectrum observed at the Earth; it is clear from Fig. 1 that θ_{obs} may be less than θ_{scat} but it cannot be greater. By the obvious analogue of equation (4)

$$\Delta\lambda/\lambda \leqslant \lambda/(z\theta^2_{obs})$$

where z is the distance of the screen from the Earth. Hence

$$v\tau = \lambda/\theta_{obs} \geqslant (z\Delta\lambda)^{\frac{1}{2}} \qquad (7)$$

Thus, with $\Delta\lambda = 5$ cm, $z = 1·5 \times 10^{20}$ cm, and $v < 30$ km s^{-1} we should have $v\tau \geqslant 3 \times 10^5$ km, or $\tau > 10^4$ s.

A similar argument leads to the same result when motion of the source is considered. z then represents the distance from the screen to the source in relation (7). With a screen close enough to the source, and perhaps associated with it, one might then hope to account for short-period variations. As shown by the alternative scales on Fig. 3, scintillations can still be achieved with tolerable values of ΔN and a for $L = z = 1·5 \times 10^{16}$ cm, giving $\tau \simeq 100$ s if other parameters in relation (7) are left unchanged. One cannot account for still shorter variations (such as pulse-to-pulse variations), however, for the finite size of the source now becomes important. The argument is simple and general: suppose that a point source must move a distance x to shift the observed diffraction pattern by its own correlation length; then the time scale of scintillations is given by $v\tau = x$. But if the source consists of incoherent elements spread out over a region of diameter $d > x$, the simultaneous presence of their displaced diffraction patterns smoothes out any possible scintillations. The time scale of scintillation must thus exceed the time it takes the source to move by its own diameter. For a source of 6,000 km diameter (white dwarf?) and $v = 30$ km s^{-1}, $\tau > 200$ s. To obtain pulse-to-pulse variations we should require a source on the scale of 10 km with a screen at a distance not much greater than 10^{11} cm containing (compare with Fig. 3 and relations (5)) 10–100 km irregularities with electron density fluctuations of 10–100 cm^{-3}; or else a very large relative velocity between source and screen. (Note that most kinds of orbital motion are already excluded, for the corresponding displacements in pulse times have not been observed[1].)

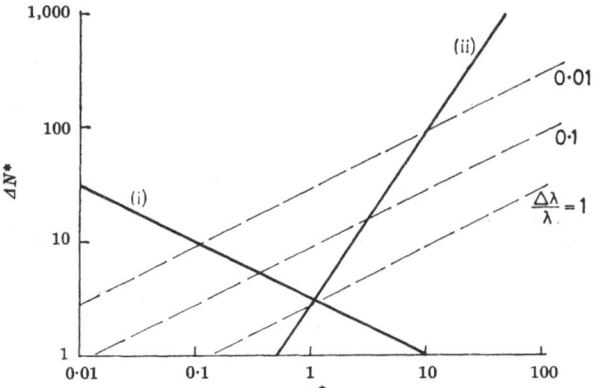

Fig. 2. The conditions required for large amplitude fluctuations, in terms of the scaled parameters ΔN^* and a^*. The point representing the screen must lie above lines (i) and (ii); the bandwidth over which scintillations are similar is shown by the dashed lines. Note that deep scintillation necessarily implies $(\Delta\lambda/\lambda) < 1$.

Amplitude Distribution of Pulses

When a plane wave is scattered by a weakly scattering screen ($\Delta\varphi \ll 1$) the diffracted field amplitude has a Rice distribution. For thicker screens ($\Delta\varphi > 1$) the amplitude distribution approximates to a Rayleigh distribution, and consequently the distribution of intensity I becomes exponential.

$$d(\text{probability}) = \exp\left(-I/I_0\right) dI/I_0 \qquad (8)$$

The distribution of individual pulse heights from CP 1919 at 81·5 MHz has been plotted by Collins (private communication) and is not exponential, having a much longer tail of very large pulses. The same appears to be true of CP 0950. So far as longer-term variations are concerned, an adequate statistical sample is not yet available to me, but the observations to date suggest that here, too, the high intensity tail of the distribution is considerably larger than in an exponential distribution. Over intermediate periods, in which there is no correlation between successive intervals, the distribution of the running means necessarily approaches a Gaussian distribution about a mean.

Though the amplitude distributions are inconsistent with scattering by a single screen, it should be noted that long-tailed amplitude distributions could arise from more complex scintillation processes. As a very simplified example, consider a screen (with $\Delta\varphi > 1$) close to the source, which projects a diffraction pattern on to a second screen (also with $\Delta\varphi > 1$) further away from the source, and suppose also that all that part of the second screen which contributes to the radiation observed by us lies within one correlation length of the diffraction pattern from the first screen. The modulation of the observed intensity is then the product of two exponentially distributed random variables. The result is easily shown to be a distribution of the form

$$d(\text{probability}) = K_0(\sqrt{2I/I_0}) dI/I_0 \qquad (9)$$

where K_0 is a modified Bessel function. The function (9) is not (and is not expected to be) a quantitative fit to the data, but it shows the required qualitative features, that is, much larger proportions of very large and very small intensities than an exponential distribution.

Discussion

Pulse-to-pulse variations are probably not caused by a scintillation mechanism, because:

(a) There are severe difficulties with the time scale. The screen would have to be fairly dense and close to the source, and the source would have to be either a neutron star or a source of highly directional radiation to have a sufficiently small effective size.

(b) It has been reported that pulse-to-pulse variations are closely correlated over a wide range of frequencies[3]. Fig. 2 shows that this is incompatible with deep scintillations.

(c) The distribution of pulse amplitudes has a much longer tail than an exponential distribution.

Scintillation is a much more attractive explanation for the long-period variations, but there are severe difficulties in this case too:

(a) Time scale. The 1 min variations in CP 1919 at 81·5 MHz, and the variations observed in three sources at 151 MHz with time scales of a few minutes[3], are too fast to be explained in terms of the general interstellar medium, unless the source moves with a speed of several thousand km s^{-1}.

(b) A screen which produces strong scintillation ($\Delta\varphi > \pi$) at 922 MHz[3] has $\lambda/(z\theta^2_{\text{scat}}) < 1$ at this frequency, and therefore has $\Delta\lambda/\lambda = \lambda/(z\theta^2_{\text{scat}}) < (81\cdot5/922)^3$ at 81·5 MHz; that is, scintillations near 81·5 MHz become uncorrelated within 80 kHz. Large amplitude variations are observed with a receiver bandwidth of 1 MHz, however, and indeed in the case of CP 0834 the amplitude

is coherent over a range of 4 MHz near 80 MHz[7]. Thus no diffracting screen can account for both the high frequency and the low frequency variations.

The question still arises whether $\Delta\lambda/\lambda$ is necessarily so small at low frequencies if high and low frequency scintillations are produced by different regions of the interstellar medium. It turns out that a consistent model can be devised, but only if one screen is very much closer to the source than the other. Suppose screen 1, producing deep scintillation at a short wavelength λ^+, is at distance z_1 from the source, and screen 2, at distance $z_2 > z_1$ from the source and z from the observer, is capable of producing deep scintillations at a long wavelength λ. The radiation falling on screen 2 has an angular spectrum of width $(z_1/z_2)\theta_{\text{scat1}}$, and will behave just like an incoherent source with this angular diameter if we observe with a receiver the bandwidth of which exceeds $\Delta\lambda/\lambda$ at the long wavelength. This angular diameter will blur out the scintillation caused by screen 2 unless

$$(z_1/z_2)\theta_{\text{scat1}}z < (\text{scale of diffraction pattern at observer})$$

$$\simeq \lambda\Big/\left\{\left(\frac{z_1}{z_1+z}\right)\theta_{\text{scat 2}}\right\}$$

Combining this with the conditions that screens 1 and 2 can produce deep scintillations at wavelengths λ^+ and λ, respectively, we find that

$$\frac{z_1}{z+z_2} < \frac{z_2}{z}\left(\frac{\lambda^+}{\lambda}\right)^3 \qquad (10)$$

Another type of model is also possible. Suppose screen 1 produces the long wave scintillations (and has little effect at the short wavelength λ^+), while screen 2 produces the scintillation at wavelength λ^+. The long wave scintillations caused by screen 2 have $(\Delta\lambda/\lambda) < (\lambda^+/\lambda)^3$ and will be smoothed out by any normal receiver bandwidth, but the amplitude variations caused by screen 1 will be observed provided the scale of the diffraction pattern falling on screen 2 is larger than that part of screen 2 contributing to the radiation observed at any one time: that is, if

$$\frac{\lambda}{(z_1/z_2)\theta_{\text{scat1}}} > z\frac{z_2}{z+z_2}\theta_{\text{scat2}}$$

Taken together with the conditions for deep scintillation at each screen, this leads again to the condition (10) obtained for the first model.

The second of these models seems the more plausible as an approximation to a real situation. For screen 2 we might take the general interstellar medium, which could easily produce scintillations at frequencies of the order of 1,000 MHz with time scales of some hours, as observed[3]. Screen 2 would then have to be the medium within 0·01 to 0·1 pc of the source to account for scintillations at frequencies around 100 MHz. The smaller distance of the screen from the source would permit the shorter time scale of these variations, and the strength of the irregularities required (see Fig. 3), though a good deal greater than that of the general interstellar medium, would not be implausibly great. If this model is in any sense correct, then additional amplitude variations should be found at low frequencies when the sources are observed with very narrow-band receivers.

I thank Dr P. F. Scott, Mr R. A. Collins and Miss J. A. Bailey for permission to use their observations before publication, and for useful discussions.

Received May 24, 1968.

[1] Hewish, A., Bell, S. J., Pilkington, J. D. H., Scott, P. F., and Collins, R. A., *Nature*, 217, 709 (1968), (Paper 1).
[2] Pilkington, J. D. H., Hewish, A., Bell, S. J., and Cole, T. W., *Nature*, 218, 126 (1968), (Paper 2).
[3] Lyne, A. G., and Rickett, B. J., *Nature*, 218, 326 (1968), (Paper 10).
[4] Uscinski, B. J., *Phil. Trans. Roy. Soc.*, A, 262, 609 (1968).
[5] Little, L. T., *Plan. Spa. Sci.* (in the press).
[6] Bramley, E. N., *Proc. Roy. Soc.*, A, 225, 515 (1954).
[7] Scott, P. F., and Collins, R. A., *Nature*, 218, 230 (1968), (Paper 8).

Applications

47. Measurement of the Interstellar Magnetic Field

by

F. G. SMITH

University of Manchester,
Nuffield Radio Astronomy
Laboratories, Jodrell Bank

THE discovery of linear polarization of the radio emission from the pulsating radio stars[1] provides an opportunity for the first direct measurement of the average magnetic field over distances of up to 100 pc from the Sun. It has already been shown[2,3] that the frequency dispersion in arrival time of the pulses can be entirely explained by a dispersion in group velocity in ionized interstellar gas, and for the pulsating star CP 1919 the integrated electron density $\int N\,dl$ along the line of sight has been found[3] to be 12·55 cm^{-3} pc. If a magnetic field has a component H_{11} along the line of sight, then the Faraday rotation of the plane of polarization is a measure of $\int N\,H_{11}\,dl$; the ratio of the two integrals then gives $\bar{H}_{11}$, the average value of H_{11} along the line of sight, weighted according to the electron density.

The Faraday rotation may be measured by observing the difference in position angle of linear polarization in adjacent frequency bands. The total rotation θ, in rad, along a line of sight is given by $\theta = 7\cdot3 \times 10^{10}\,\nu^{-2}\int N\,H_{11}\,dl$, where ν is the frequency in MHz, H_{11} is in G, N in cm^{-3}, and l in pc. For CP 1919, for example, the rotation at the radio astronomy frequency band of 151·5 MHz is then $40\cdot7 \times 10^6\,H_{11}$ rad, and the rate of change of θ is $0\cdot54 \times 10^6\,H_{11}$ rad MHz^{-1}.

CP 1919 is a particularly interesting source for this measurement, because it lies at galactic co-ordinates $l=55°$, $b=+3°$, very close to the direction of the local spiral arm of the Galaxy. Unfortunately the linear polarization of this source is not complete, and it is probably variable. Measurements have therefore been made on the source CP 0950 which, in contrast with CP 1919, emits very strong pulses with practically complete linear polarization at a position angle which appears to remain nearly constant for considerable periods. This source is at $l=229°$, $b=43°$, which is at about 45° to the spiral arm; it is also the source with the least dispersion in arrival

time, and presumably one of the closest pulsating stars, possibly at a distance of only 10–20 pc. The measurements of magnetic field presented here therefore refer to a comparatively local part of the Galaxy.

Observations were made with two receivers at 151·5 MHz, using a bandwidth of 1 MHz, connected to orthogonal plane polarized dipoles in the Mark I telescope. It was found that ionospheric Faraday rotation occasionally oriented the plane of polarization of the pulses parallel to one dipole, and that the ratio of pulse intensity in the two receivers could then exceed 20 : 1. It may be shown that the differential Faraday rotation across the receiver bandwidth must be less than 0·2 rad for such a high degree of polarization to be observed even if the source is completely polarized. For CP 0950 the measurements of dispersion in arrival time give a value of $\int N\,dl = 3\cdot02$ cm^{-3} pc leading to a rate of change $d\theta/d\nu = 0\cdot13 \times 10^6\,H_{11}$ rad MHz^{-1}. The value of $\bar{H}_{11}$ must therefore be less than $1\cdot6 \times 10^{-6}$ G.

Two further receiver channels, centred at 147·5 MHz and 149·5 MHz, each with 1 MHz bandwidth, were then also connected to one of the dipole feeds, and recordings of a series of individual pulses were made with a fast galvanometer, as in Fig. 1. From 1930 to 2100 UT on April 3, 1968, the relative strength of the recorded pulses changed over a period of 1·5 h by an amount consistent with the change in Faraday rotation during the evening decline of the ionosphere. (The source was near the meridian at this time, at an elevation of about 45°, so that the line of sight component of the Earth's field was not changing.) The relative sensitivity of the four channels to the pulses was unknown, but no changes in receiver gain were made during the experiment.

Pulses of moderate height, avoiding effects of receiver saturation, were chosen from recordings made every few minutes. The pulse heights were plotted in Fig. 2, after

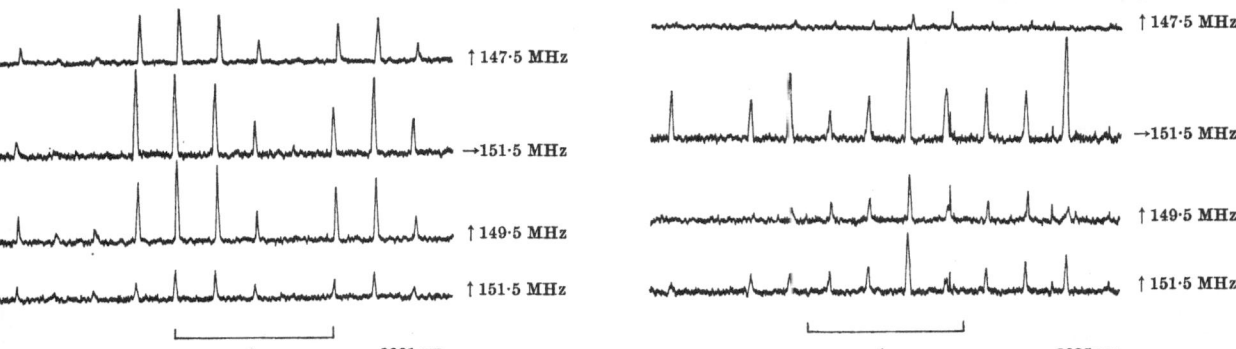

Fig. 1. Pulses from CP 1919 recorded on linearly polarized receiver channels at 147·5, 149·5 and 151·5 MHz, at 2001 UT and 2035 UT on April 3, 1968.

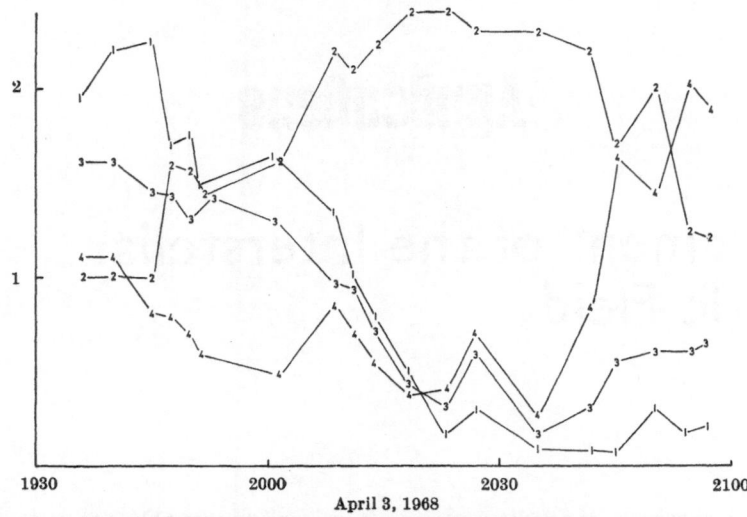

Fig. 2. Relative pulse intensities on the four receiver channels at 147·5 MHz (1),
149·5 MHz (3) and 151·5 MHz (4), and on the orthogonal polarization at 151·5 MHz (2).

applying an arbitrary scaling factor to each channel and an appropriate factor to each pulse so as to preserve a standard pulse intensity.

Channels 2 and 4 were receivers at the same frequency, connected to orthogonal dipoles. As the ionospheric Faraday rotation decreased during the evening, the pulse strength reached a maximum in channel 2, and passed through zero on channel 4, as would be expected. Channels 1 and 3 at frequencies 4 MHz and 2 MHz below channel 4, but connected to the same dipole, also pass through zero at nearly the same time as channel 4, as would be expected if the Faraday rotation changed very little with frequency. The relative gains of the channels apparently changed somewhat during the experiment, so that the angle of rotation between channels 1, 3 and 4 cannot be determined accurately, but it is evident that a zero pulse strength was recorded in the order channels 4, 3 and 1, with a time interval of about 4 min between each channel. This value has been used for an estimate of the rotation between channels by noting that the ionospheric rotation was at the rate of about 90° h⁻¹, as seen on channels 2 and 4.

In this way the rotation is estimated to be 0·065 rad MHz⁻¹. The total rotation at 150 MHz is therefore about 4 rad, which is comparable with the total rotation expected from the ionosphere in the circumstances of this experi-ment. Furthermore, the observation that the lower frequency channels passed through zero later than the higher frequency channel, while the ionospheric Faraday rotation was decreasing, makes it clear that the measured total rotation is in the same sense as the ionospheric contribution. The conclusion is that the interstellar Faraday rotation is less than 2 rad at 150 MHz, so that the weighted mean value of the field is less than 2×10^{-7} G.

This field is very much less than existing estimates of the general galactic field, but it should be noted that the measurement refers to one particular line of sight in a local part of the Galaxy. It is also remarkable that there can be no appreciable Faraday rotation within the source of the radiation, even though there must exist there a magnetic field large enough to generate the impulsive radiation.

I would like to acknowledge the collaboration of J. G. Davies, B. J. Rickett and A. G. Lyne in these observa-tions.

Received April 11, 1968.

[1] Lyne, A. G., and Smith, F. G., *Nature*, **218**, 124 (1968), (Paper 7).
[2] Hewish, A., Bell, S. J., Pilkington, J. D. H., Scott, P. F., and Collins, R. A., *Nature*, **217**, 709 (1968), (Paper 1).
[3] Davies, J. G., Horton, P. W., Lyne, A. G., Rickett, B. J., and Smith, F. G., *Nature*, **217**, 910 (1968), (Paper 6).

48. Interstellar Magnetic Field

by

R. S. ROGER

Dominion Radio Astrophysical Observatory,
Penticton, B.C., Canada

W. L. H. SHUTER

Department of Physics,
University of British Columbia,
Vancouver

SMITH[1] has given an upper limit to the interstellar magnetic field in the direction of the pulsating radio source CP 0950 derived from measurements of its Faraday rotation. This limit, $< 2 \times 10^{-7}$ gauss for a component directed towards the Earth, is, as Smith points out, much less than other existing estimates. Furthermore, it is significant that the contribution to the total rotation from the Earth's iono-sphere implied by this limit is considerably lower than one would expect for the local time of the observations. In this communication we suggest that the observations as described are compatible with an interstellar field component of about 10^{-6} gauss directed towards the source and with a larger rotation contribution from the ionosphere.

It is likely that the critical frequency (f_0F2) on the ray path at the time of the observations ($\sim 20^h$ UT April 3, 1968) was about 7·6 MHz (private communication from T. R. Hartz). This, combined with a probable F-region equivalent thickness of 305 km (ref. 2), suggests a total electron content of about 22×10^{12} cm^{-2}. At an elevation of 45° on the meridian at Jodrell Bank, one would expect about 12 radians of ionospheric Faraday rotation at 150 MHz.

Evidence that the interstellar field is directed towards CP 0950 ($l = 229°$, $b = 43°$) and is therefore oppositely directed to the geomagnetic field over Jodrell Bank is available from measurements of the rotation measure (RM) of polarized radiation from the radio galaxy $3C$ 227 ($l = 229°$, $b = 42°$). This is given by Morris and Berge[3] as RM$= -6 \pm 3$ radians m^{-2} and by Gardner and Davies[4] as RM$= -7 \pm 3$ radians m^{-2}. From these data we estimate that for CP 0950 at a distance of ~ 30 pc (ref. 5) the

expected rotation measure will be $-6 \cdot 5 \times \dfrac{30}{140} = -1 \cdot 4$ radian m^{-2}, where we have assumed the path in the galactic disk in which Faraday rotation takes place for $3C$ 227 as 140 pc. This implies an interstellar Faraday rotation at 150 MHz of $-5 \cdot 6 \pm 1 \cdot 8$ radians for CP 0950.

It seems reasonable to conclude that the total rotation of $+4$ radians observed by Smith could be made up of $+12$ radians in the ionosphere and -8 radians in interstellar space. This interpretation is consistent with the observation[1] that the total rotation was in the same sense as the ionospheric contribution and implies an interstellar magnetic field four times larger than the upper limit set by Smith. Further measurements of the total Faraday rotation at a time of day when the ionospheric contribution is considerably smaller would be sufficient to decide which of the two interpretations is correct.

Received May 7, 1968.

[1] Smith, F. G., *Nature*, **218**, 325 (1968), (Paper 47).
[2] Roger, R. S., *J. Atmos. Terr. Phys.*, **26**, 475 (1964).
[3] Morris, D., and Berge, G. L., *Astrophys. J.*, **139**, 1388 (1964).
[4] Gardner, F. F., and Davies, R. D., *Austral. J. Phys.*, **19**, 129 (1966).
[5] Lyne, A. G., and Rickett, B. J., *Nature*, **218**, 326 (1968), (Paper 10).

49. General Relativistic Influence on Observed Pulsar Frequencies if Pulsars are Orbiting Objects

by

BANESH HOFFMANN

Queens College,
City University of New York,
Flushing, New York

RECENTLY[1], I pointed out that pulsars might be a means of making a new test of the general theory of relativity by providing a stable extra-terrestrial frequency standard against which the differential general relativistic effect on terrestrial clock rates at perihelion and aphelion could be measured.

The purpose of this communication is to point out what could be an important analogous effect should the pulsars prove to be orbiting bodies.

Suppose the pulsars were orbiting bodies, and for simplicity consider the case of a pulsar the orbit of which is in a plane perpendicular to the line of sight. The contribution of its motion to the first order Doppler effect will then be essentially zero, and one might be tempted to assume that once the contribution of the Earth's orbital motion had been subtracted, any residual variation in the observed frequency of recurrence of the pulses would arise from a variation in the basic pulse rhythm of the source itself. As was previously pointed out[1], however, there would be a small general relativistic effect of about 1 part in 2×10^9 arising from the ellipticity of the terrestrial orbit. In addition to this, as will be shown, there could be a non-annual residual variation in the observed frequency that did not arise from a variation in the intrinsic pulse rhythm of the source itself, provided the orbit of the source was significantly different from a circle.

Let the maximum and minimum orbital distances of the source from its star be R_a and R_p, respectively, let the mass of the star be M, and let the intervals between pulses measured in Schwarzschild co-ordinate time at

maximum and minimum orbital distances from the central star be dt_a and dt_p.

Then[2],

$$dt_p/dt_a - 1 = 3MG(R_a - R_p)/2c^2R_aR_p$$

where G is the Newtonian gravitational constant and c is the speed of light. For the Earth this comes to $4 \cdot 9 \times 10^{-10}$.

It could be smaller or larger for an orbiting pulsar source depending on the values of R_a, R_p and M. Because the Schwarzschild coordinates used in deriving this formula are static, the effect will be transmitted faithfully to the Earth and will appear as a residual variation in the rhythm of the received pulses. Because the effect does not depend on the orientation of the plane of the pulsar orbit relative to the line of sight, it will yield information involving the eccentricity of the orbit, the mass of the central star and the period of the orbital motion, even though the orbit is here assumed to be in a plane perpendicular to the line of sight.

Should pulsars prove to be orbiting objects, there could be a problem concerning their intrinsic periodicities if the above effect, which could be relatively large for some of them, were not taken into account in analysing the data. Moreover, if the problem were removed by taking account of this effect we would have yet another confirmation of general relativity.

Received April 5, 1968.

[1] Hoffmann, B., *Nature*, **218**, 667 (1968).
[2] Hoffmann, B., and Sproull, W. T., *Amer. J. Phys.*, **29**, 640 (1961).

50. Pulsars and a Possible New Test of General Relativity

by
BANESH HOFFMANN
Queens College,
City University of New York,
Flushing, New York

THE recent discovery of rapidly pulsating radio sources[1] with a remarkably uniform periodicity suggests the possibility of testing a general relativistic effect involving the orbital motion of the Earth and the gravitational field of the Sun.

Hoffmann and Sproull[2] have pointed out that the difference in the general relativistic effect on terrestrial clocks of the solar gravitational potential at perihelion and aphelion amounted to a frequency shift of order $3 \cdot 3 \times 10^{-10}$. Furthermore, this frequency shift, far from being essentially nullified by the relativistic centrifugal effect at perihelion and aphelion, was actually augmented, so that a total general relativistic relative frequency shift of order $4 \cdot 9 \times 10^{-10}$ could be expected after the first order Doppler effect had been subtracted.

While such a frequency shift is large compared with the accuracy of modern atomic clocks, there seemed no way to detect it, because to do so an extra-terrestrial frequency standard was needed, against which to compare the frequencies of terrestrial clocks over a period of six months. The only reasonable extra-terrestrial standards were spectral frequencies of solar or other stellar atoms, and the sharpness of the spectral lines was inadequate.

With the discovery of the pulsars, the situation may have changed. Already, Hewish et al., in their first report[1], have remarked that by observations extending for a year it should be possible to test the constancy of N_0 to about one part in 3×10^8. This corresponds to an accuracy of 3×10^{-9}, which differs from the general relativistic effect of 5×10^{-10} by only about an order of magnitude.

One can reasonably expect increased precision in the measurement of the N_0 of pulsars, and also a shortening of the time period over which measurements must be made to achieve this increased precision. It is also possible that other pulsars will be discovered that have even greater uniformity of periodicity than do those already known. The discovery of pulsars therefore raises the possibility of a new test of the general theory of relativity.

In principle, the method would be to look for residual periodic variations, having a period of one year, in the apparent N_0 of pulsars, N_0 being apparently greatest at perihelion and least at aphelion and the difference amounting to about 1 part in 2×10^9. The interpretation would be that the pulsars had stable frequencies and that the terrestrial clocks also had stable frequencies, but that the solar curvature of space–time and the centrifugal effect of the Earth's orbital motion gave rise to an annually varying "red-shift" that made terrestrial clocks, when compared with the pulsar rhythms by means of light signals, appear to go more slowly at perihelion than at aphelion.

Note added in proof. Drake[3] has indicated a present accuracy of one part in 10^8 in ten days with the likelihood of increasing this by perhaps two orders of magnitude. The experiment thus appears feasible.

Received April 2, 1968.

[1] Hewish, A., Bell, S. J., Pilkington, J. D. H., Scott, P. F., and Collins, R. A. *Nature*, 217, 709 (1968), (Paper 1).
[2] Hoffmann, B., and Sproull, W. T., *Amer. J. Phys.*, 29, 640 (1961).
[3] Drake, F. D., *Science*, 160, 416 (1968).

51. Interstellar Magnetic Field

by
F. G. SMITH
University of Manchester,
Nuffield Radio Astronomical Laboratories,
Jodrell Bank

Roger and Shuter[1] have pointed out that the recent measurement (Smith[2]) of the interstellar magnetic field from the Faraday rotation of the plane polarized radio waves from the pulsar CP 0950 may be interpreted as a combination of ionospheric and interstellar rotation. They conclude that the upper limit of 2×10^{-7} gauss for the interstellar field should be replaced by an actual value of about 10^{-6} gauss.

This interpretation is based on values of the ionospheric electron density obtained only from estimates of the penetration frequency at the time of observation. The actual values of penetration frequency at the Radio and Space Research Station, Slough, are available through the kindness of the director, and they show that the new

interpretation is not justified. From 18h to 22h UT on April 3, 1968, the penetration frequency fell linearly from 7 MHz to 4·8 MHz. Roger and Shuter quote a value of 7·6 MHz at 20h UT. Our present estimate of the ionospheric contribution is less than theirs by a factor of just over two chiefly because of this difference. Their calculation must therefore be revised to read: Measured rotation, +4 radians; ionospheric contribution, +6 radians; interstellar contribution, −2 radians. The original conclusion is unchanged, and the upper limit of the interstellar magnetic field remains at 2×10^{-7} gauss.

Received June 17, 1968.

[1] Roger, R. S., and Shuter, W. L. H., *Nature*, 218, 1036 (1968), (Paper 48).
[2] Smith, F. G., *Nature*, 218, 325 (1968).